Urban Sustainability

The Urban Sustainability Book Series is a valuable resource for sustainability and urban-related education and research. It offers an inter-disciplinary platform covering all four areas of practice, policy, education, research, and their nexus. The publications in this series are related to critical areas of sustainability, urban studies, planning, and urban geography.

This book series aims to put together cutting-edge research findings linked to the overarching field of urban sustainability. The scope and nature of the topic are broad and interdisciplinary and bring together various associated disciplines from sustainable development, environmental sciences, urbanism, etc. With many advanced research findings in the field, there is a need to put together various discussions and contributions on specific sustainability fields, covering a good range of topics on sustainable development, sustainable urbanism, and urban sustainability. Despite the broad range of issues, we note the importance of practical and policy-oriented directions, extending the literature and directions and pathways towards achieving urban sustainability.

The series will appeal to urbanists, geographers, planners, engineers, architects, governmental authorities, policymakers, researchers of all levels, and to all of those interested in a wide-ranging overview of urban sustainability and its associated fields. The series includes monographs and edited volumes, covering a range of topics under the urban sustainability topic, which can also be used for teaching materials.

Tianyao Zhang

Towards Healthy Settlements

Health Implications of Residential Suburbanization in Guangzhou

Springer

Tianyao Zhang
Department of Urban Planning and Design,
School of Architecture
Harbin Institute of Technology
Shenzhen, Guangdong, China

ISSN 2731-6483 ISSN 2731-6491 (electronic)
Urban Sustainability
ISBN 978-981-97-1206-9 ISBN 978-981-97-1207-6 (eBook)
https://doi.org/10.1007/978-981-97-1207-6

This Springer imprint is published by the registered company Springer Nature Singapore Pte Ltd.
The registered company address is: 152 Beach Road, #21-01/04 Gateway East, Singapore 189721, Singapore

Acknowledgments

This book derives from my doctoral dissertation, which owes much to the supports and help of many people around me. First and foremost, I'd like to express my sincere gratitude to my doctoral supervisor, Prof. Rebecca L. H. Chiu, for the patient and inspired guidance and mentorship she provided to me during the entire period of my research. It has been an honor to be her Ph.D. student. I appreciate all her contributions of time and ideas to make my Ph.D. experience productive and stimulating. Her immense knowledge, rigorous academic attitude, and the enthusiasm she has for the research have been contagious and motivational for me, and she has provided me an excellent example as a successful woman scholar and professor. Prof. Chiu is not only a great academic advisor but also a life mentor to me, who has offered invaluable advices that would benefit all my life. In addition, many thanks also to the Department of Urban Planning and Design and the University of Hong Kong for providing me with the opportunity, financial support, and nice environment and valuable resources to conduct this study.

I am grateful to the academics and professionals who have provided me great supports in data collection and analyses. I am sincerely thankful to Dr. Yang Yu in Department of Housing and Urban-Rural Development of Guangdong Province and Mr. Lou Yanfeng in Huayang International Engineering Design Limited Company, who helped to organize and participated in the interviews. Special thanks to Dr. Huang Dingxi, Mr. Zhou Wen, Mr. Cai Chun for their professional opinions on the planning regulations and institutional environment of urban development in Guangzhou. Many thanks to my schoolmates graduated from South China University of Technology for their assistance and supports during the field trip, especially to Mr. Chen Ruimin, Ms. Chen Ye, Ms. Ling Jingya, Ms. Li Ting, Mr. Ma Xing, Ms.Wu Bijun, Mr. Xiao Jingkun, Ms. Ye Suiting, Ms. Zhang Jinlin, and Ms. Zhong Minyi. In addition, I would like to express my gratitude to the housing managers and directors of residents' committee from the three neighborhoods for their participation in the interviews; special thanks to Mr. Huang Chao, who is the director of the residents' committee of AG, for his great help in recruitment of the interviewees and the provision of a nice place. I would also like to acknowledge Mr. Qiu Linsheng and his six colleagues for conducting the questionnaire survey with me in the case areas in summer time.

Furthermore, I am hugely indebted to Prof. John Bacon-Shone in the University of Hong Kong for their generous support and suggestions on the statistical analysis. Last but not least, I am thoroughly grateful to Prof. Zhang Chunyang, Prof. Liu Yuting, Dr. Wei Lihua, and Dr. Deng Zhaohua in South China University of Technology for their great suggestions and encouragement.

Additionally, I gratefully appreciate all my colleagues and friends in HKU who offered their supports and care, encouraging and helping me to get through all the tough periods during the four years. Special thanks to Sun Yi, Wang Jiejing, Zhang Xianchun, and Zhang Xiaohu, who gave me generous suggestions and recommendations on this study, and many thanks must go to Guo Mengdi, Liu Dongyuan, Ma Yiguan, Song Jing, Wang Xi, and Xu Yueli, with whom the wonderful time were spent together.

Finally, I would like to express my heartfelt gratitude to my family for all their unconditional love and encouragement. My parents have been always giving me their great supports in my life, which is the essential spiritual force for me to keep on moving. In addition, my husband, Mr. Xia Sheng, whom I deeply love, has been the most important supporter through the four years, and I am greatly appreciate to his forever love, valuable advices, infinite patience, and supports. In particular, many thanks to my best girlfriend Ms. Meng Lingyu, who has known me for more than 20 years, for her ongoing accompany and sagacious comfort in my life. Finally, I would like to dedicate this book to all my family members and my best friends.

Praise for *Towards Healthy Settlements*

"This excellent book identifies the key factors that influence residential behavior and health outcomes in urban suburbs of China through in-depth case study analysis. Based on the framework of human ecology theory, it reveals the interaction between environmental and behavioral factors in terms of health effects, and attempts to explain the institutional factors of residential environment factors that affect health in housing development. The rigorous analytical framework and in-depth case studies of this book provide valuable insights for scholars, policy makers, and practitioners who are interested in understanding the health effects of housing in the process of urban suburbanization and housing development in China."

—Suhong Zhou, *Professor, Former Vice Dean, School of Geography and Planning, Sun Yat-sen University; Director, Guangdong Provincial Engineering Research Center for Public Security and Disaster*

"With a comprehensive analytical framework between built environment and health outcome, this excellent book provides a solid study on health impact of the suburban commodity housing estates (SCHEs). By offering both qualitative and quantitative analyses, Dr. Tianyao Zhang reveals the complicated health implications of residential suburbanization in the context of South China. This book contributes significantly to healthy city literature by unraveling the association among health performance and its multiple influential factors related to suburbanization. It is an important reference for scholars and practitioners in the field of urban planning, health geography and public health."

—Lan Wang, *Chair Professor, Healthy City Science and Planning, College of Architecture and Urban Planning, Tongji University*

Contents

Abbreviations

AG	Agile Garden estate
BMI	Body mass index
GV	Asian Games Village
HOA	Homeowner association
LNT	Luoxi New Town estate
OD	Odds ratio
RMD	Risks of mental disorders
SCHEs	Suburban Commodity Housing Estates
WHO	World Health Organization

Chapter 1
Introduction

Urban health is believable to be a considerable challenge of the twenty-first century [11]. The population booms in most countries after the Second World War accelerated the progress of urbanization, after which the relationship between urbanization and health have gradually become apparent in the academic literature [11]. The importance of health for urban development has been increasingly recognized, in particular with the revision of globally sustainable development goals, which propose that the equity, inclusiveness and accountability in health and development are the core principles and themes of sustainable development [43]. Being seen as both a key determinant and a resource, health is increasingly highlighted as a central component of sustainable development [9, 21, 36]. To capture the benefits of health for sustainable development, a considerable body of knowledge has been generated in various fields focusing on the health effects of urbanization, such as environmental health, infectious disease public health, and lifestyle-related behavioral health [11]. Specifically, urbanization and the resulting changes towards sedentary lifestyles lead to a remarkable transformation in the problematization of health, shifting from infectious to chronic disease etiology [35]. However, the causation or correlation between health and urbanization have not been explained by the unequivocal and empirically validated theories, particularly it is difficult to ascertain the causal relations within specific local contexts [11, 19, 37]. In response to these constrains, an integrated, multi-disciplinary, multi-sectoral approach for urban health research is necessary, one that integrates urban planning, social psychology, epidemiology, public health and environmental sciences [10].

Since the 1980s, stimulated by economic, land and housing reforms, China has experienced rapid urbanization and a rising urban population, the scale of which is unique from elsewhere in the world. Subsequently, commodity housing boomed, extending to the suburban areas of Chinese cities. Accompanying with both urban spatial restructuring and socioeconomic transition, lifestyle has changed dramatically, incurring health consequences to China's urban population. Consistent with the global trend, there has been an epidemiological transition from infectious diseases

T. Zhang, *Towards Healthy Settlements*, Urban Sustainability,
https://doi.org/10.1007/978-981-97-1207-6_1

to chronic diseases in China during a short period, with the focus of public health shifting from life expectancy to health expectancy [10]. Attention is now being paid to the health implications of urban development in China, given that the transition to western-style diets and sedentary lifestyles, the outdoor air pollution, and the prevention of accidents are significant factors affecting the health of China's population in future [20]. However, the government, communities and planners still do not fully understand the health consequence of environmental and lifestyle changes [37].

Thus, to depict the global–local nexus for this study, this chapter explores the global discourse on health and urbanization, highlighting the rising awareness of health in urban studies, and the healthy cities approach is then reviewed as an intervention for urban health and sustainability worldwide. Considering the specific economic and social transformation, this chapter examines the health tolls of urban development in China, especially for residential suburbanization. Based on the above review, this chapter states the research aim, research questions and the research scope of this study.

1.1 Global Discourse on Health and Urbanization

1.1.1 Urbanization, Urban Living and Health Concerns

Urban living has become a reality for the majority of the world's population [40]. Expeditiously urbanization is one of the complicated and significant socio-economic phenomena of the twentieth and twenty-first centuries, transforming the rural society to urban one and changing the production and consumption procedures [1, 19]. According to the WHO [43] the world's urban proportion is projected to "increase from 54% in 2015 to 60% in 2030 and to 66% by 2050," with the most growth of urban population in developing countries, particularly in Asia and Africa [40]. This is consistent with the fact that the rate of urban growth in developing countries is much higher due to the economic engines [41]. Urbanization has made urban living to become a predominant way of living for most of the world's population, under the context of which the health implications of transforming to urban living are widely discussed.

Historically, cities have been associated with the evolution of ideas and practice of public health [31], and urban living has been considered to be associated with health both positively and negatively [40]. Due to urbanization, cities are obviously at the forefront of tackling with global development issues and account for a high ratio of both burden and causes of public health issues [43] Especially, cities in the low- and middle-income countries face severe health threats caused by the conflicts between the rapidly growing urban populations and diminishing resources [43].

The threats for health for city residents have been changing with the dynamic socio-economic and environmental contexts of urban living, leading to the transformation in disease etiology. The 2016 WHO report on urban health considers the urban lifestyle a driving factor for the spread of non-communicable diseases, estimated to have caused 60% of global deaths in 2012 [43]. The main health threats in cities have traditionally been diarrheal disease and respiratory infections stemming from crowding because of urbanization, but now it is the physicochemical hazards derived from urban environment such as air pollution, traffic hazards, urban heatwaves, and climate change [31]. In addition, health disparities have been found between developed and developing countries in different periods. In developed countries, after the middle of the twentieth century, the increasing health disparities between urban and suburban, rural population lie in the higher prevalence of HIV epidemic and violence in urban area [40]. While in developing countries, apart from the popularity of infectious diseases in cities in the twentieth century, the growing urban middle-class had seen a rise in chronic diseases in the twenty-first century [40].

To summarize, as a predominant social context globally, urban living has been inevitably associated with the risk factors or determinants of population health [19]. Overall, "a rising prevalence of non-communicable diseases, the persistent threat of infectious disease outbreaks and an increased risk of violence and injuries are key public health concerns in urban areas" [43, p. 19].

1.1.2 Rising Discourse on Health in Urban Studies

1.1.2.1 Relink of Urban Planning and Public Health: Urban Health Studies

Until the beginning of the twenty-first century, the literature on the characteristics of modern urban areas and the aspects of urban living that affecting human health has been sparse [40]. This may be due to the involvement of multiple disciplines such as urban planning, urban sociology and epidemiology and the lack of collaboration among them [40]. However, promoted by the relink of the two fields of urban planning and public health, health and urban professionals are now paying particular attention to the mechanisms of urban environment affecting human health, increasing the discourses of health determinants in the urban context.

The connections between urban planning and public health had been under ups and downs in the twentieth century. In the middle of nineteenth century, rapid industrialization resulted in the squalid living conditions in cities, giving rise to the cooperation of modern urban planning and public health [3, 31, 32]. However, ties between these two fields frayed with the improvement of urban living conditions, and the focus of urban planning shifted to the macro socioeconomic and environmental issues while that of the public health to the micro level studies at the first half of the twentieth century [8, 12, 29]. Nevertheless, the links between urban planning and public health have been restored since the 1980s. With the substantial influence of urbanization on health, urban health has come forth as a sub-discipline of international public health since the mid-1980s [26]. With the Ottawa Charter for Health Promotion in 1986, the era of "new public health" has begun [5, 26, 46]. Unlike traditional public health, the new public health draws from social science in its concern for health promotion, effects of social capital and physical environment on lifestyle, chronic diseases and mental disorders, social equity, community participation, and inter-departmental cooperation [5, 26, 46]. This focus transformation of health promotion has led to the collaboration of urban planning and public health to a large extent.

As an emerging sub-discipline, urban health studies focus on the channels of the characteristics of urban environment affecting human health [19]. As an interdisciplinary field, many studies from multi-disciplines, such as epidemiologic, public health, health geography, and urban planning try to explore the connection between urban environment and health [35]. Several studies have examined the effects of the built environments on "physical activity, asthma, obesity, cardiovascular disease, lung cancer mortality, and mental health" [37]. The mechanism of disease spreading in urban context is another key point for urban health studies, in addition to three broad categories of theories and mechanisms are referred to: the physical and the social environment, the availability of and accessibility to health and social services [19, 41]. However, the causal relationship between the urban environment and specific diseases is difficult to ascertain, implying the pressing need of identifying the mechanisms by which urban environment positively and negatively affects health [37]. Overall, the key challenges in urban health lie in: (1) the inconsistent definitions and uses of urbanization and urban, (2) the inadequate illustration discussion on how and why the urban context may influence health, (3) the complexity of causation in urban context [19]. Thus, greater attention to urban health is called for.

1.1.2.2 Health and Sustainable Development

Another factor accelerating the discourse of health in urban studies is the promoting of sustainable development, to achieve which health is crucial as the other side of the same coin of urban sustainability [24]. In terms of political agenda, the first principle of Rio Declaration (1992) states that "people are at the center of concerns for sustainable development and that they are entitled to a healthy and productive life in harmony with nature" [22]. WHO has placed health centrally on sustainable development agenda, where health is one of five crucial elements[1] that has been paid special attention to [36]. Additionally, in the adopted set of 17 global development goals (Sustainable Development Goals[2]) in 2015, the target of making cities inclusive, safe, sustainable and resilient has been proposed, and its achievement depends on the important determinants of human health in cities such as housing, air quality and transport simultaneously [43]. Thus, the importance of health with reference to sustainable development has been recognized as the central component of human development [9, 36]. Although health emphasizes reducing deaths and disease while

[1] The other four aspects include: water, energy, biodiversity, and agriculture.

[2] "In September 2015, 193 Member States of the United Nations gathered in New York City, USA, to adopt a new set of 17 global development goals with 169 national targets called the Sustainable Development Goals (SDGs)" [43, p. 14].

sustainable development highlights the environmental impact, a quantity of synergies is probably to arise in the achievement of the two goals indicating the opportunities for producing co-benefits for both health development and urban development [43]. Hancock [25] argues that healthy communities must be environmentally and socially sustainable, since the global change stemming from the industrial economy influence the web of life and the quality of built and natural environment is determinative to human health.

In such a context, the research and practice of public health and sustainable development are interlinked, taking account of urbanization centrally [30]. In addition, the links between various urban environmental issues and health have become the focus of many studies [36]. Evidence shows that poor quality environments and social and economic disadvantages are against health, and the health risks almost cover all dimensions of sustainability [33]. Thus, health and sustainability are mutually decided: "a more sustainable society cannot be anything other than a healthier society" [33], sustainable development could not be achieved with a high prevalence of debilitating illness and poverty, and human health could not be sustained without intact life-support systems and healthy environments [36]. So, it is important to consider the health dimension in the pursuit of sustainable development, and providing for a healthy world population without undermining sustainability is the major challenge in the twenty-first century [22].

1.1.3 Healthy Cities Approach: An Intervention for Urban Health and Sustainable Development

In the hope of reintegrating the fields of urban planning and public health, healthy cities approach has been gradually regarded as an intervention to achieve urban health and sustainable development globally [3]. In the field of new public health, health promotion is revived for the healthy cities approach, beginning with the Ottawa Charter for Health Promotion [23]. In 1986, the WHO healthy cities project was established as a global movement, providing a local basis for translating WHO principles into a tangible global program to promote health [2, 4]. According to the definition given by the WHO, the purpose of healthy cities approach is to improve the physical and social environments and well-being of the people who live and work in urban areas, and to strengthen community resources that enable people to support each other and to achieve their maximum potential [2, 23, 42]. With this holistic well-being aim, healthy cities approach is multidimensional, including social, political, environmental, economic and cultural aspects, as well as urban planning, human services, heritage, community empowerment and participation [23]. In this context, healthy cities approach is a holistic political and urban development framework for global health promotion.

Promoted by the increasing implementation of healthy cities projects, researchers begin to pay special attention to health intervention in urban studies. The healthy cities approach seeks to "put health on the agenda of decision makers in cities and develop a local, participatory approach to dealing with health and environmental problems" [42]. To support the healthy cities approach, healthy urban planning has been highlighted in terms of shaping the conditions that have both positive and negative effects on human health [13]. The principles of healthy cities approach that have implications for urban planning are summarized as: equity, sustainability, intersectoral cooperation, community involvement, and international action and solidarity [4]. Conceptually, the idea of "urban health niche", a framework for public health-focused urban planning, has been conceptualized by [35], and they propose a systemic healthy city planning model accordingly, highlighting three coexisting systems where multilevel spatial determinants of health could be identified. The three systems are individual-population system, household-neighborhood-city system, and governance and decision-making systems, of which components could be connected by the interactive causal process, determining the health of an individual fundamentally [35].

Furthermore, referring to the experience of the US in the early twentieth century, the concept of healthy city could see its origin in the aspiration of local governments of establishing the conditions for health [23]. The rationale of the healthy cities approach is based on the good urban governance model, involving "broad political commitment, intersectoral planning, citywide partnerships, community participation, monitoring and evaluation" [2]. Good urban governance is crucial under the advancement of capitalist globalization, accompanied by which the twin crises are challenging the implementation of healthy cities approach, namely, ecological unsustainability and social class polarization [2]. These crises have deleterious effects on the human health of developing countries in particular [2]. Thus, an effective implementation of healthy cities approach is severely challenged by the neoliberal ideology denoting powerful economic and political interests. It is criticized that "practical, evidence-based, context-specific interventions that can improve the health of the majority of the world's city-dwellers are more important than public health shibboleths" [2]. This also applies to China, which has been undergoing rapid mercerization in the past three decades.

1.2 Health and Suburbanization in China: The Local Context

1.2.1 Health Tolls of Urban Development

1.2.1.1 Background: Urbanization, Suburbanization, Housing Development

In China, the economic, land and housing reforms from 1980s have transformed the urban environment dramatically, evidenced by cities proliferation, urban expansion and decentralization [17, 45]. Rapid urbanizing since 1990s has led to increase in urban population from 191 million in 1980, to 622 million in 2009 and an expected additional 292 million people between 2014 and 2050 [20, 28, 43]. With this speedy urbanization, cities have to provide additional housing. As a result, there has been a boom of commodity housing estates with the deepening of housing reform and the maturing of housing market; there has been urban expansion in the form of residential suburbanization; and studies of population redistribution and land-use change demonstrate that suburban districts are emerging as the new growth poles [15, 49, 50]. Thus, accompanied by housing development, urbanization in China is distinguished from other developing countries by the absence of widespread proliferation of urban slums, and this is partly because of the China's governmental commitment to expand housing and basic infrastructure in suburban area [20].

Additionally, another unique characteristic of China's urbanization is the simultaneous processes of urbanization and suburbanization. Suburbanization could be seen as decentralized urbanization, and it is argued that Chinese suburbs are the core element of urbanization [7, 14]. In terms of the initiatives, suburbanization has changed from a government-led mode in the 1980s to a market-led mode with the boom of commodity housing development after the 1990s [15, 51]. On the whole, suburbs in China are obligate for three functions: receiving relocation population generated by urban renewal and reconstruction in the central cities since 1980s, providing luxury villas and gated communities for a class of car-driving homeowners, and settling massive waves of migrants from other regions [51]. Particularly, residential suburbanization is widespread in China's large cities, driven and made possible by rising car ownership and preference for suburban quality and spacious housing since market mechanisms gradually dominates [15]. Thus, the development mechanism, physical characteristics and socioeconomic status of the suburbs of China are all distinctive from those of Western countries. In short, the galloping urbanization and economic affluence call for living space larger and better, extending the metropolitan area into rural area, creating suburbanization and changing the lifestyles of the urban population in China.

1.2.1.2 Health Consequences of Urbanization in China

As discussed, there is an undeniable relation between urbanization and health. Studies conducted on China have consistently found that urbanization and urban spacial expansion contribute to the changes of land use, urban environment and residents' lifestyle, affecting human health problems accordingly [28]. Rapid urbanization and suburbanization have created sprawl, dispersive urban functions, heavy reliance on automobiles, inefficiency of service facilities, and low walkability, as well as various environmental impacts, such as the depletion of natural and rural land, inefficient resources consumption, air pollution, and waste [18]. In this context, the health consequences of the ongoing urbanization are of utmost importance for China's population [20]. There are both positive effects and adverse effects of urbanization, on one hand, China's urban spatial expansion is closely associated with economic development, with urbanization driven by and driving economic growth [20, 28], contributing to the improvement of urban living conditions, nutrition and health care system, declining mortality and infectious disease and increases in life expectancy [28, 39]. On the other hand, China's urban population is continuously exposed to risk factors for disease, in particular the ones related to the challenging environmental and social conditions, such as environmental pollution and physical inactivity [20]. The spread of environmental, economic and social changes brought by urbanization potentially increase the prevalence of major risk factors for chronic disease at the same time [39].

With rapid urbanization and economic growth, an epidemiological transition occurred in China in only a few decades, shifting the leading causes of mortality from the infectious disease and perinatal conditions to chronic diseases and injuries [47]. For example, the prevalence of hypertension has soared since 1980s, and the proportion of overweight and obese population has experienced rapid growth by 39% and 97% respectively from 1992 to 2002 [47]. The increases prevalence of obesity and hypertension is most obvious amongst males, urban and high-income groups [39]. The causes of the transition to chronic non-communicable diseases mainly lie in societal change, lifestyle choices and behavioral factors [43]. The adverse changes in lifestyle (such as a high intake of dietary fat and increased physical inactivity) accompanying industrialization and urbanization could increase the risk of chronic disease [27]. The pace and spread of behavioral changes, such as changing diets, decreased physical activity, and high rates of male smoking, has expedited to a remarkable degree [47].

Taking the increasing prevalence of overweight and obesity as an example, it is mainly because of reduced physical activity in the context of the little change in the average daily calorie intaken for Chinese population over the past two decades [47, 52]. As a result, there is a growing burden in urban areas in terms of chronic diseases, preventable morbidity and mortality, and the associated health-care costs [47]. With vascular disease and cancer becoming the leading causes of death among Chinese adults [27], diseases like cardiovascular, diabetes, cancer, chronic respiratory diseases and mental disorders has estimated to cost China around US$27.8 trillion during 2012–2030 [43]. This makes the necessity of the intervention to promote healthy lifestyle, such as the strategies for controlling hypertension, smoking cessation, increasing physical activity, and improving nutrition, which contribute to reduce the burden of premature death among adults in China [27].

1.2.1.3 Lifestyle Changes and "New Middle-Class" in China

Considering the crucial effects of urban lifestyle on human health, the lifestyle changes in urban China are a key point for this study. Lifestyle is determined by the housing choices to a large extent, which are deeply embedded in the sociocultural and spatial reconfiguration of Chinese society [16]. Physically, with rapid urban growth triggered by urbanization, urban decentralization is emerging and suburban areas have become a growth anchor in China, where commodity housing development has been significant. Socio-culturally, the affluent residents (urban new middle-class) have developed specific ideas about their living environment and lifestyle, responding to which the advertising slogan "to choose a house means to choose a lifestyle" has appeared in housing market, implying that residential compounds could be regarded as the basis for lifestyle formation [16].

Stimulated by the economic reforms, a high consumer-oriented professional middle-class has emerged in China, promoted by the strategy of redistribution of public assets, especially of housing [38]. As a result, a middle-income oriented housing reform has been stimulated, and the urban environment has been restructured through deliberating re-zoning of the urban space, resulting in "the emergence of gated communities and segregated residential neighborhoods populated by new homeowners" [38]. Unlike the "masses", the "urban new middle-class" aspires homes with more green space, better air quality, and spaciousness but privacy and exclusivity [16]. The emergence of the "new middle-class" matches with the target population of the expanding supply of packed and branded suburban residential compounds, which create a new middle-class living [44, 48]. This could be regarded as a newly emerged urban class with a distinctive lifestyle and therefore a unique set of health implications.

In terms of the definition of middle-class, focusing on the contributions that the middle class makes, scholars usually link the middle class with democracy, entrepreneurs, human capital, and consumption. In defining the middle class, both relative and absolute terms are employed by existing studies, mainly focusing on the annual income or consumption capability, which is arbitrary to some extent [34]. Based on the purpose of this research, the 13 middle class is defined as the population who are well-educated, with non-manual work, especially professional occupations, and with median annual income ranging from USD 10,000 to USD 60,000.[3] Focusing on the role of the middle class in promoting urban development in China, it could be identified with the gated housing development, employment, consumption patterns, family socialization and lifestyles which have produced a distinctive social fabric for

[3] Based on the criteria given by "Forbes".

the class formation. In this sense, it is curious to know the health conditions of this urban middle class and the unique set of health implications of the suburban living.

1.2.2 Initial Stage of Healthy Cities Intervention

Regarding the health consequence of urbanization and urban living, the polity and practice interventions in China are still in quite an early stage although some significant tempts have been made. The idea of healthy cities was introduced to China in 1994, and several cities such as Suzhou and Shanghai were chosen for pilot tests. According to the "Twelfth Five-Year Plan for economic and social development of Guangzhou" (2011), Guangzhou has been planned to develop into a healthy city. After two years of preparation, the first "Construction Plan for Healthy City of Guangzhou (2011–2020)" was launched in January 2013. This plan aims to improve urban infrastructure and environmental quality, to promote the development of ecological and healthy industry, and to create a harmonious relationship among people, resources and environment, ensuring a healthy and sustainable living environment.[4] Additionally, five health objectives are proposed in this plan: constructing a healthy environment, building a healthy society, improving health service, nurturing a healthy population, developing healthy industry. Finally, this plan proposes fifteen health actions, covering urban environmental quality, pollution control, healthy community construction, health education, promotion of physical activity, and chronic disease prevention and control.

However, the document itself and the reports of its implementation suggest that this healthy city plan of Guangzhou is rhetoric. First, the planning department is responsible for constructing a healthy environment by making and implementing urban plans and improving infrastructures.[5] It is a very general description of the tasks of planning department in a broad context, which might be difficulty to implement in practice. In fact, no great attention to health has been instituted in the planning and design process, and no effective guidance and supervision from upper-level plans and official authorities respectively have been issued for the planning and design. Furthermore, although it is acknowledged that the healthy cities approach should be implemented through interdepartmental cooperation, few studies and practices have been undertaken to establish a good urban governance model for building a healthy environment. The existing administrative structures of Chinese government could pose obstacles to the implementation of a healthy city policy and plan.

In terms of urban planning and design, the following concepts concerning health have emerged since the 1980s: comprehensive development mode, people-centrism, comfortable housing design (which evolved into healthy housing design), and sustainable development. In housing development, sustainable and people-oriented

---

[4] Source: http://www.gz.gov.cn/gzgov/s9132/201302/1585890.shtml.

[5] http://www.gz.gov.cn/gzgov/s9132/201302/1585890.shtml.

housing design appeared in 2001, and "Techniques for construction healthy housing" were issued in 2001 and revised in 2002 and 2004. Under this guideline, the physical environmental health and social environmental health gained considerable attentions in housing planning and design [6]. This technical specification proposed the criteria for constructing healthy residential and social environment, ranging from indoor physical comfort and outdoor healthy requirements. However, although it is comprehensive guidance for healthy housing construction, the effectiveness of this guidance is limited mainly because of the fact that it is only a set of professional guidelines with no legal status.

1.2.3 Research Gaps

Mounting evidence suggests that human health relates to urbanization process both globally and locally. Health disparities persist between urban and rural regions, especially with a higher incidence of chronic diseases and a lower degree of physical activity in urban areas [39, 53]. More attention should be paid to the possible health effects of the dynamic urban spatial expansion in terms of various changes in urban environment at different development phases, not just simply to a broad urban–rural comparison [28]. In addition, the inconsistent definition of urbanity exacerbates the difficulties in filling the research gap, and this situation could be much remarkable as suburbanization in China is recognized in association with urbanization.

Empirical studies have been attentive to the health consequence of vulnerable migrants in rapidly urbanizing China, but there is no interest in the emerging "new middle-class", who is more affluent and enjoys higher socioeconomic status. Additionally, the lifestyle of "new middle-class" changes due to suburban living, especially in terms of travel behavior and physical activity. The changes bear inconstant implication to their health conditions, which have been neglected in the exiting literature.

Conceptually, consistent with the challenges pointed out in urban health studies, the causal mechanisms for the urban context influencing human health are still debatable, especially in different local contexts. Although many studies have suggested the relations between physical and mental health issues and human-modified environment including homes, schools, workplaces, parks, industrial areas, and roads, the effects of mediating and moderating factors need further exploration.

Furthermore, although the urban environmental quality and the ways of managing and using by its inhabitants are indispensable to human health and well-being, few urban planning systems incorporate health issues [4]. This remains true in China because of the initial stage of health cities intervention in urban development. For the implementing of healthy city plans, current studies in China have paid little attention to the importance of amenable planning institution for creating healthy living environments. In addition, concerning the need for interdepartmental collaboration for implementing the healthy cities approach, the empirically-oriented, evidence-based, and context-specific studies guided by robust theoretical foundations on urban governance are necessary.

1.3 Research Aim and Research Questions

1.3.1 Research Aim

In China, it is believed that suburbanization is spearheaded by meeting the housing demands of the speedily growing "new middle-class". This new type of housing is usually organized in the form of housing estate with common facilities, and it is important to know whether the suburban housing is providing a healthy environment for its residents. Given the lack of research on the health effects of suburban living in China, this study aims to formulate policy and planning recommendations for achieving a healthy neighborhood living environment for the middle-income residents of suburban commodity housing estates by underlying the health determinants stemmed from suburban housing development both quantitatively and qualitatively. To achieve this aim, two important issues should be highlighted: the causal relations between health performance and the suburban living and the underlying institutional causes that create the suburban living environment. Accordingly, three research questions could be raised to facilitate achieving the research aim, covering the health performance evaluation, health influential factors identification, and examination of the institutional mechanism in housing development process, as detailed in the next section.

1.3.2 Research Questions

(1) *What is the health performance of residents in the suburban commodity housing estates (SCHEs) of Guangzhou?*

As discussed, the development of SCHEs plays an important role in China's urban spatial restructuring and in nurturing a distinctive kind of suburban lifestyle to meet the aspiration of the emerging middle-income population. To understand health and suburban living in China in its global and local contexts, the health performance of residents living in SCHEs should be evaluated. With the epidemiological transition to chronic diseases, the evaluation on the health performance is conducted in the dimensions of physical and mental health.

(2) *What and how have the factors stemming in suburban living affected the health performance of residents?*

As both a result and a cause of suburbanization, SCHEs has formulated a distinctive living environment in China, one whose health implications have been overlooked. Based on the evaluation results of the residents' health, the second question explores whether the SCHEs offer a health-promoting residential environment and help to induce the health-facilitating residential behaviors. Two tasks will be undertaken to answer this question. First, identify the factors that influence health by investigating the suburban residential living from the perspectives of the residential environment and health-related behaviors. Second, to explain the ways of the factors affecting the health performance, with special attentions given to the interactions of residents with the residential environment.

(3) *How have the institutional factors within housing development created the neighborhood living environment?*

It is necessary to explore the institutional factors that determine the characteristics of suburban residential living related to the health performance with regards to SCHEs development. The residential environment and residential behaviors are ultimately shaped by housing development. The roles and relations among different stakeholders within housing development process are scrutinized, including the local government, developers, and planners at inception, decision-making, the planning and design stages, and the housing managers, residents' committees, and ordinary residents at the housing operation stage.

By seeking answers to the three questions, the effects of suburban residential living on the health of their residents could be understood, and the influence of the housing development process on residents' health, especially the effects of institutional factors, would be better captured, adding knowledge to China's health intervention in the housing development process.

1.4 Significance of This Study

This study will provide a critical understanding of the health implications of suburban living in the unique context of urbanization and urban expansion (suburbanization) in China. Notably, the outcomes of this study will generate significant knowledge on the nature of SCHEs in Guangzhou in relation to residents' health performance (especially the general physical health, incidence of obesity, depression, and risks of mental disorders). Empirically, this research will deepen the knowledge of the health effects of suburban living on residents' health performance in terms of both residential environment and health-related residential behaviors. In addition, the daily routine nurtured by SCHEs and its health effects will be unfolded, broadening the scope of urban health literature, which has been grounded in developed Western countries. In particular, the investigation of a unique suburbanization and the health consequence of "new middle-class" population in a fast-developing country will add a new dimension to urban health literature.

Conceptually, by building upon the existing discourse on urban environment and health, this study will expand the application of the people-environment transaction modes from the lens of social ecology in understanding the causation between environmental and behavioral factors and human health. In addition, this research will enrich the use of institutional mechanisms to explain with reference to housing development the formation of specific residential environment and the chain reaction on the health performance.

In terms of policy implementation, the findings of this study will enable urban academic and government officials to understand the health effects of urban development and the effectiveness of health interventions in public policies. Ultimately, this study will inspire institutional innovation towards good urban governance for health in future suburban development, contributing to better choices regarding urban spatial expansion and housing development strategies, residential planning and design, and innovations of suburban development.

1.5 Research Scope

This study focuses on the dynamics of residential suburbanization in Guangzhou since the launch of housing marketization reforms in late 1990s. Stimulated by the establishment and maturation of socialist land and housing markets, residential suburbanization features the proliferation of housing estates, leading to urban spatial expansion and restructuring and decentralization. By referring to the different socioeconomic development contexts, three stages of residential suburbanization in Guangzhou are identified and included in the study, and three representative SCHEs are selected as the study cases. All are located in Panyu district, with increasing distance from the urban core area. For each case, both the physical components and the subjective perception of the residential environment and the health-related residential behaviors are the focuses of the investigation. In addition, the institutional factors in commodity housing development will be examined as an explanation to the effects of institutional mechanism on the residential environment directly and residents' health indirectly.

1.6 Outline of the Book

The book is comprised of eight chapters. The first four chapters introduce the research background, presenting the conceptual and empirical review and research methodology. Afterwards, Chaps. 5–7 answer the three research questions, and based on the study findings, conclusions and policy implications are presented in the final chapter.

This introductory chapter highlights the global discourse on health in relation to urbanization and the health tolls of urban development, with special reference to suburbanization in China. Based on this review, research aim and questions are formulated. Chapter 2 establishes the conceptual foundation for this study by discussing the concepts of suburbs, suburbanization, and healthy neighborhoods, examining the health determinants at neighborhood scale and the institutional factors in housing development affecting the formation of residential environment. Chapter 3 explores the issues of SCHEs development in light of healthy neighborhoods, with a special focus on the scrutinization on the three stages of development of suburban housing in Guangzhou. In addition, the health implications of suburban living in China are examined, especially for the threats of obesity and mental disorders. Chapter 4 explains the research methodology, consisting of the analytical framework, data collection and analysis methods.

Chapters 5 through 7 are the main body of analyses that focus on the three research questions. Chapter 5 is comprised of two sections of analyses; first, it assesses the health performance of residents living in the three selected SCHEs. In the second section, it describes the socio-demographic status of the three cases as control variables, followed by the analysis on the physical health in relation to the physical activity. Chapter 6 identifies the environmental and behavioral factors for the mental health performance based on regression analyses, and the effects of residential environment and health-related residential behaviors on mental health are examined both qualitatively and quantitatively. Chapter 7 consists two sections with the purpose of explaining underlying factors affecting the creation of a healthy living environment. In the first section, residents-environmental transactions in SCHEs are examined, based on which the obstacles in SCHEs for achieving the healthy living environment are discussed. The second section illustrates the institutional factors that formulate the residential environment during the suburban housing development, which influence residents' health accordingly. Finally, Chap. 8 concludes, presenting the main findings and their broader implications on the creation of a healthy urban living environment in China.

1.7 Conclusion

This chapter presents the background, themes and significance of this study, based on which the research aim and questions are established. It reviews the global discourse on health and urbanization, including the trend of revived interest in urban planning and public health, the rising studies of urban health, and the relationship between health and sustainable development, highlighting the importance of health concerns in urban development. Then the specific urbanization especially the residential suburbanization in China is reviewed, followed by the review of health consequences of urban spatial expansion and lifestyle changes. Based on the reviews in current global–local contexts, the research gaps are identified, contributing to the establishment of research aim and research questions. Finally, the outline of the book is presented according to the framework for addressing the three research questions.

References

1. Allen A (2009) Sustainable cities or sustainable urbanisation. Palette UCL's J Sustain Cities 1(2)
2. Awofeso N (2003) The healthy cities approach: reflections on a framework for improving global health. Bull World Health Organ 81(3):222–223
3. Barton H, Grant M, Mitcham C, Tsourou C (2009) Healthy urban planning in European cities. Health Promotion Int 24(suppl_1):i91–i99. https://doi.org/10.1093/heapro/dap059
4. Barton H, Tsourou C (2000) Healthy urban planning: a WHO guide to planning for people. Taylor & Francis
5. Baum F (2003) The new public health. Oxford University Press
6. Chu X (2002) An study on the concepts and tendencies of the development and design of city housing at housing-merchandisize period in China [Matser dissertation]. Tsinghua University
7. Chung CJ, Inaba J, Koolhaas R, Leong ST (2001) Great leap forward: Harvard design school project on the City. Taschen GmbH
8. Corburn J (2009) Toward the healthy city: people, places, and the politics of urban planning. MIT Press
9. Corvalán CF, Kjellström T, Smith KR (1999) Health, environment and sustainable development: identifying links and indicators to promote action. Epidemiology 10(5):656
10. de Hollander AEM, Staatsen BAM (2003) Health, environment and quality of life: an epidemiological perspective on urban development. Landsc Urban Plan 65(1):53–62. https://doi.org/10.1016/S0169-2046(02)00237-2
11. de Leeuw E (2001) Global and local (glocal) health: the WHO healthy cities programme. Global Change Human Heal 2(1):34–45. https://doi.org/10.1023/A:1011991014805
12. Ding G, Cai J (2013) Public health and planning: a discussions on health impact assessment and tools for urban and rural planning. Urban Plann Forum 5:48–55
13. Duhl LJ, Sanchez AK (1999) Healthy cities and the city planning process: a background document on links between health and urban planning (EUR/ICP/CHDV 03 04 03). WHO Regional Office for Europe. https://apps.who.int/iris/handle/10665/108252
14. Feng J (2001) The progress and prospect of the research on suburbanization in China. Hum Geogr 16(6):30–35
15. Feng J, Zhou Y, Wu F (2008) New trends of suburbanization in Beijing since 1990: from government-led to market-oriented. Reg Stud 42(1):83–99. https://doi.org/10.1080/00343400701654160
16. Fleischer F (2007) To choose a house means to choose a lifestyle. The consumption of housing and class-structuration in Urban China. City & Society 19(2):287–311
17. Fleischer F (2010) Suburban Beijing: housing and consumption in contemporary China. University of Minnesota Press
18. Frumkin H, Frank LD, Jackson RJ (2004) Urban sprawl and public health: designing, planning, and building for healthy communities. Island Press
19. Galea S, Vlahov D (2005) Urban health: evidence, challenges, and directions. Annu Rev Public Health 26(1):341–365. https://doi.org/10.1146/annurev.publhealth.26.021304.144708
20. Gong P, Liang S, Carlton EJ, Jiang Q, Wu J, Wang L, Remais JV (2012) Urbanisation and health in China. The Lancet 379(9818):843–852. https://doi.org/10.1016/S0140-6736(11)61878-3
21. Goodland R (1995) The concept of environmental sustainability. Annu Rev Ecol Syst 26(1):1–24. https://doi.org/10.1146/annurev.es.26.110195.000245
22. Haines A, Alleyne G, Kickbusch I, Dora C (2012) From the Earth Summit to Rio+20: integration of health and sustainable development. The Lancet 379(9832):2189–2197. https://doi.org/10.1016/S0140-6736(12)60779-X
23. Hancock T (1993) The evolution, impact and significance of the health cities/healthy communities movement. J Public Health Policy 14(1):5–18. https://doi.org/10.2307/3342823
24. Hancock T (1996) Health and sustainability in the urban environment. Environ Impact Assess Rev 16(4):259–277. https://doi.org/10.1016/S0195-9255(96)00024-8

25. Hancock T (2000) Healthy communities must also be sustainable communities. Public Health Rep 115(2–3):151–156
26. Harpham T, Molyneux C (2001) Urban health in developing countries: a review. Prog Dev Stud 1(2):113–137. https://doi.org/10.1177/146499340100100202
27. He J, Gu D, Wu X, Reynolds K, Duan X, Yao C, Wang J, Chen C-S, Chen J, Wildman RP, Klag MJ, Whelton PK (2005) Major causes of death among men and women in China. N Engl J Med 353(11):1124–1134. https://doi.org/10.1056/NEJMsa050467
28. Li X, Wang C, Zhang G, Xiao L, Dixon J (2012) Urbanisation and human health in China: spatial features and a systemic perspective. Environ Sci Pollut Res 19(5):1375–1384. https://doi.org/10.1007/s11356-011-0718-7
29. Li Y, Zhang X, Xia C (2023) Towards a greening city: how does regional cooperation promote urban green space in the Guangdong-Hong Kong-Macau Greater Bay area? Urban Forest Urban Green 86:128033
30. McCarthy M (1999) Sustainable urban development-the public health dimension. J Environ Sci (China) 11(3):279–283
31. McMichael AJ (2000) The urban environment and health in a world of increasing globalization: issues for developing countries. Bull World Health Organ 78(9):1117–1126
32. Northridge ME (2003) Sorting out the connections between the built environment and health: a conceptual framework for navigating pathways and planning healthy cities. J Urban Health: Bull New York Acad Med 80(4):556–568. https://doi.org/10.1093/jurban/jtg064
33. Porritt J (2005) Healthy environment—healthy people: The links between sustainable development and health. Public Health 119(11):952–953. https://doi.org/10.1016/j.puhe.2005.08.004
34. Ravallion M (2009) Decentralizing eligibility for a federal antipoverty program: a case study for China. The World Bank Econ Rev 23(1):1–30. https://doi.org/10.1093/wber/lhn023
35. Sarkar C, Webster C, Gallacher J (2014) Healthy cities: public health through urban planning. Edward Elgar Publishing
36. von Schirnding Y (2002) Health and sustainable development: can we rise to the challenge? The Lancet 360(9333):632–637. https://doi.org/10.1016/S0140-6736(02)09777-5
37. Srinivasan S, O'Fallon LR, Dearry A (2003) Creating healthy communities, healthy homes, healthy people: initiating a research agenda on the built environment and public health. Am J Public Health 93(9):1446–1450. https://doi.org/10.2105/AJPH.93.9.1446
38. Tomba L (2004) Creating an urban middle class: social engineering in Beijing. China J 51:1–26. https://doi.org/10.2307/3182144
39. Van de Poel E, O'Donnell O, Van Doorslaer E (2009) Urbanization and the spread of diseases of affluence in China. Econ Hum Biol 7(2):200–216. https://doi.org/10.1016/j.ehb.2009.05.004
40. Vlahov D, Galea S (2003) Urban health: a new discipline. The Lancet 362(9390):1091–1092. https://doi.org/10.1016/S0140-6736(03)14499-6
41. Vlahov D, Freudenberg N, Proietti F, Ompad D, Quinn A, Nandi V, Galea S (2007) Urban as a determinant of health. J Urban Health 84(1):16–26. https://doi.org/10.1007/s11524-007-9169-3
42. Werna E, Harpham T, Blue I, Goldstein G (1998) Healthy city projects in developing countries: an international approach to local problems. Earthscan
43. WHO UH (2016) Global report on urban health: equitable, healthier cities for sustainable development. World Health Organization, Geneva, Switzerland
44. Wu F (2010) Gated and packaged suburbia: packaging and branding Chinese suburban residential development. Cities 27(5):385–396. https://doi.org/10.1016/j.cities.2010.06.003
45. Wu F, Xu J, Yeh AG-O (2006) Urban development in post-reform China: state, market, and space. Routledge
46. Xie X (2006) Introduction of new public health. Henan J Preventive Med 17(2):128–128
47. Yang G, Kong L, Zhao W, Wan X, Zhai Y, Chen LC, Koplan JP (2008) Emergence of chronic non-communicable diseases in China. The Lancet 372(9650):1697–1705. https://doi.org/10.1016/S0140-6736(08)61366-5

48. Zhang X, Cheung DM, Sun Y, Tan J (2019) Political decentralization and the path dependent characteristics of the state authoritarianism: an integrated conceptual framework to understand China's territorial fragmentation. Eurasian Geogr Econ 60(5):548–581
49. Zhang X, Chen Y, Xu Y, Yue W, Tang C (2023a) Does regional cooperation constrain urban sprawl? evidence from the Guangdong-Hong Kong-Macao Greater Bay Area. Landsc Urban Plan 235:104742
50. Zhang X, Shen J, Sun Y, Zhou C, Yang Y (2023b) Formation of city regions from the bottom-up initiatives: investigating coalitional developmentalism in the Pearl River delta. Ann Am Assoc Geogr 113(3):700–716
51. Zhou Y, Logan JR (2008) Growth on the edge: the new Chinese metropolis. In: Urban China in transition. John Wiley & Sons, Ltd., pp 140–160. https://doi.org/10.1002/9780470712870.ch6
52. Zhou C, Chan RCK, Zhang X (2023) District government as a policy entrepreneur: understanding policy experimentation in China's Wujin district. J Urban Affairs, 1–23.https://doi.org/10.1080/07352166.2023.2208757
53. Zhu Y-G, Ioannidis JPA, Li H, Jones KC, Martin FL (2011) Understanding and harnessing the health effects of rapid urbanization in China. Environ Sci Technol 45(12):5099–5104. https://doi.org/10.1021/es2004254

Chapter 2
Suburbanization and Healthy Neighborhood: A Conceptual Review

To comprehend the health implications of suburbanization in China, this chapter intends to lay a theoretical foundation by reviewing the literature related to suburbanization and health-promoting environments. A human ecological perspective is adopted, focusing on the transactions between people and their environments. Suburbanization as a distinctive form of urban expansion and residential neighborhoods as identifiable human settlements are selected as the basis for analysis. In addition, a healthy neighborhood is treated as a theoretical concept as well as a goal of urban development. A human ecological conceptual framework for understanding health determinates in the urban environment is formulated, and institutional factors in suburbanization and housing development are raised as an explanatory tool, which formulate the residential environment and the resultant way of life.

This chapter comprises three main sections. The first section presents the concepts and definitions of suburbs and suburbanization in the contexts of West and China. The second section focuses on concepts related to health-promoting environment, including health, healthy neighborhood, and people-environment transactions. The literature on the interaction between residents and the built environment in relation to health promotion is reviewed, and consequently a human ecological conceptual framework for understanding health-promoting environment is formulated. In the third section, important health determinants at neighborhood scale are explored, focusing on environmental and personal factors, in addition to the institutional factors in China's suburbanization and housing development.

T. Zhang, *Towards Healthy Settlements*, Urban Sustainability,
https://doi.org/10.1007/978-981-97-1207-6_2

2.1 Suburbs and Suburbanization

2.1.1 *Definitions in Western Context*

Suburbs are traceable to the first cities. The prefix *sub* means beneath or below or outside, and *urbs* is the pre-urban nucleus, often fortified, sometimes a castle [44]. As places beyond city limits, suburbs have existed long before the nineteenth century, which are outskirts of towns hanging on to the central area physically and economically [94]. The usage of *suburbs* has been changing with economic, social, cultural and political development. Generally, the meaning of '*suburbs*' describes a parabolic trajectory, from low to high and then falling [44]. After the early-industrial period, the wealthy began to move from congested and unpleasant urban core areas to the distant semirural peace [44]. Although it is supposed that the first and definitive suburbs in Anglo-America were those built for and occupied by the social élite [44], there were numerous working-class suburbs in Australia and North America in nineteenth and early twentieth centuries [43, 44], and "the suburbs in European countries (with the exception of England) were often relegated to the social inferior" Harris and Larkham [44], as cited in Hohenberg and Lees 1995, pp. 306–307). Thus, the suburbs are not always associated with idyllic, quasi-rural and highly private life-style in much of the English-speaking urban tradition [44].

Globally, scholars begin to pay more attention on the diversity of suburbs across the world [15, 34, 70, 81], while the universal character of suburbanization has not been recognized to date [25, 109]. With accelerated suburbanization, increasing populations are absorbed in the expanding metropolitan regions in Africa, Asia and South America [15]. Although a typical paradigm of suburbanization has been created based on the Anglo-American model that has dominated the urban disciplines, it is supposed that a new paradigm should be developed instead of the earlier patterns in understanding the suburbanization in twenty-first century [15]. It is acknowledged that suburban development is "a consequence of social, economic and political forces specific to place and culture, one that focuses on the process of suburbanization, rather than paradigms based on earlier models" [15]. Thus, the dominant type of suburb is usually different and specific to local context, as well as a mixed feature where multiple types coexist [81]. In sum, the nature of the suburbs is influenced by the locality and by globalization.

In this context, it is very difficult to define and delimit the *suburbs* satisfactorily. There is no consensus on what elements of life and landscape should be investigated in defining *suburbs* [106, 107]. There has been a spate of the terminological and conceptual research to clarify concepts and to differentiate among commonly used terms referring to rural–urban interface, such as the suburbs, the urban fringe, the rural–urban fringe, the satellites, and the peri-urban interface [1, 71]. A relatively comprehensive study in defining suburbs is conducted by Forsyth [30], who has examined the range of suburban definitions in previous studies. According to Forsyth [30], the dimensions of the definitions of suburbs include physical, functional, social and process aspects. The sub-dimensions such as location, built environment characteristics, transportation, activities, political places, sociocultural characteristics, styles of building, time, critical assessments, and indices often appear in the definitions of suburbs [30, 110, 111]. The multi-dimensions of this concept also indicate the potential complexity in defining suburbs when different dimensions are combined [30].

2.1.2 Definitions in China

In China, defining and delimiting *suburbs* is difficult as well, and there is no comprehensive definition of *suburbs*. Originally, *suburbs* in China existed for a thousand years because there has been a tradition of city-suburbs dichotomy [38]. For instance, according to "Zhouli, Diguan, Zaishi",[1] the suburbs are beyond the city, consisting of the inner-suburbs (within 50 *li*[2] of the city) assigned to the officials, the literati, and merchants (in feudal China), and the outer-suburbs (within 100 *li* from the city), where the land was cultivated or grazed.[3]

With urban expansion and urban spatial restructuring, the definition of *suburbs* changed. In terms of the locality, the enormous influences of the economic, land and housing reforms should be referred to. Before the launch of the economic reform and opening-up policy, the urban and rural were separated absolutely by the demographic control and the ideological education based on Maoism [14]. While influenced by the reforms and the innovations, the contrast between urban and rural has become confused due to the land transformation from rural use to urban construction use and the increasing immigration to the suburbs from urban centers or other regions. Thus, the connotation of *suburbs* in China is specific to place, culture, stage of the economic development and governance.

Scholars usually define a concept depending on their substantive interests and research objects instead of defining it explicitly as a whole [30]. This is true in the definition of *suburbs* in China as well. For instance, focusing on residential development of suburbs, [107] points out that two areas that draw the most attention in defining suburbs: physical land-use composition, focusing on the intermixing of rural and urban land, and the social characteristics of the population, such as occupation, income, and lifestyle. Most attempts to define China's *suburbs* use the categorical approach to delimit the structure of metropolitan area, and the *suburbs* are often referred as a territorial entity [72].

Geographers consistently divide China's metropolitan area into three categories: urban core area, inner suburbs, and outer suburbs [29, 105, 113]. However, this categorization is arbitrary [72]. In terms of the administrative division, the municipality

---

[1] It was published in Western Zhou Dynasty, and in Chinese, it is: 周礼 • 地官 • 载师.

[2] *Li* is equal to a half kilometer.

[3] In Chinese:"邑外为郊，离城五十里为近郊，百里为远郊," "以宅田、士田、贾田任近郊之地," "以官田、牛田、赏田、牧田任远郊之地.".

(*shi*) is comprised of the city district (*shi qu*) and municipally administered county (*shi xia xian*), and the city district consists of the urban area (*cheng qu*) and the suburban area (*jiao qu*) [38]. So the "so-called inner and outer suburbs are de facto urban districts under the direct control of city governments, rather than semi- or fully independent suburbs as in Unites States" [72]. In this sense, suburbs are under the jurisdiction of the municipality, and the administrative institutions in suburbs include both the street branch office, residents' committee, town government, and village committee [38].

From the perspective of urban planning, the urban area is often divided into a built-up area, a planning area, and a suburban area. The suburban area is beyond the planning area and has a close relationship with the urban development in political, social and economic aspects [38]. Thus, it is difficult to define the boundary of suburbs due to the various criteria such as administration, socioeconomic status, history and geography [38].

Anthropologists are interested in the social characteristics of China's suburbs especially the mosaics of different social spaces coexisting in geographic proximity [72]. Suburban residents range from young, educated couples to elderly work unit employees, form urban central area to from countryside [81, 102]. In terms of urban social spaces under the label of suburbanization, they include "state-planned new towns, university towns, migrant enclaves, college graduate ghettos, artist villages, gated communities, exclusive villas, middle-class high-rise islands, resettlement housing projects, new CBDs, and ecological cities" [72, p. 356].

Thus, the literature of Chinese suburbs concentrates on their territorial and social aspects. Compared with the definitions of suburbs addressed by Forsyth [30], the definition of suburbs in China is distinctive from those in Western context. This situation could be because the "US model of suburbanization of the postwar period has been over generalized in the studies of suburbanization", which generates confusion in describing the urban reality of territorial expansions on the edge of Chinese cities [72]. While the specific types of suburbanization in China have been overlooked [72]. Thus, the locality of Chinese cities should be paid more attention to rather than applying the American model of suburbanization to understand suburbs in China, so the process of suburbanization should be emphasized in understanding suburbs rather than the paradigms of suburbanization.

Behind all the forms of suburbs are urbanization, suburbanization, and urban decentralization [25]. Diverging from the US model of suburbanization which is characterized by low density, automobile-based residential settlements, single land use, long commutes, and limited role of governments, suburbanization in China is usually featured with relatively high density, demographic diversity, transportation-oriented development, and greater 26 roles of governments [72]. In addition, unlike Western countries, Chinese suburbanization emerged after the economic reform and opening-up policy, when cities broke their traditional boundaries and generated the transitional space for residents who migrated from rural and urban core area [114]. Thus, it is argued that urbanization accompanied suburbanization in China, and suburbanization could be seen as decentralized urbanization [28]. In other words, Chinese suburbia is an outcome of urbanization [14].

After the 1990s, with lower housing price and greener environment in suburbs, more people moved to suburbs [29], indicating the process of market-led suburbanization. In addition, residential suburbanization has been driven by the growing car ownership and the preference of suburban living since 1990 [29]. Thus, stimulated by the growing influence of market mechanisms, the residential suburbanization is speeding up with the boom of suburban commodity housing estates, changing the landscape and the form of Chinese cities.

2.2 Conceptualization of Health-Promoting Environments

Suburbanization is a new pattern of urbanization, whereby housing development in the suburban areas pushes the development of urban infrastructure, facilities and finally prompts consumption its residents. Discussion of urban health is of particular importance in this trend. For SCHEs, the connotation of the healthy neighborhood helps to provide a conceptual framework to evaluate their effects on the health performance of their residents. Conceptually, the ecological model is most commonly used to assess the health of a neighborhood, and it is used as the analytical lens in this study.

2.2.1 On Health and Healthy Neighborhood

2.2.1.1 Health

The concept of health is open to various interpretations, difficult to achieve a consensus definition. "The word *health* is derived from the old English work *hal* meaning whole, healed, and sound" [55, p. 395]. World Health Organization defines health as "a state of complete physical, mental and social well-being and not merely the absence of disease or infirmity."[4] This is the most popular definition of health since many scholars acknowledge this definition in their studies [36, 51]. In addition, two levels of health could be identified in the definition: individual health and collective social well-being. When social wellbeing is considered, the interpretations of health ought to be enlarged to the collective level [55]. Stokols [84] notes out that healthfulness is a multifaceted phenomenon, comprising of physical health, emotional, spiritual and social well-being, and ranging from individual-oriented evaluation of physiological health to organizational- and community-level analyses of social cohesion. In this sense, health is a broad concept, whose determinants cover everything from DNA to global climate change [41].

Focusing on the causes of health from the human ecological perspective, Lawrence [55, p. 396] interprets health as "a condition or state of human beings resulting from

[4] Retrieved from http://www.who.int/about/definition/en/print.html.

the interrelations between humans and their biological, chemical, physical, and social environments." In addition, health is "a resultant of genetic factors, the nature and the quality of health care, behavior or lifestyle and the quality of the physical and social-cultural environment" [95]. Thus, as a basic need that enables people to lead active lives [82], health is seen as "the ability of an individual to achieve her or his potential and to respond positively to the challenges of daily life" [55, p. 396]. Thus, people-environment transactions are important for health, and people-environment congruity could be seen as the essential condition for health, emphasizing "the match between individual life satisfaction and objective standards of living ... and the extent of the possibilities for the individual to fulfill his/her goals and needs through his or her environment" [64].

Moreover, the dichotomy of physical and mental health is frequently found in existing studies. For physical health, the following often appear in the epidemiology literature: obesity, cancer and chronic disease, cardiovascular disease, respiratory disease, and sexual dysfunction [69]. Notably, numerous studies have explored the relationship between obesity and urban sprawl in the West, so one focus of this study is obesity and suburban living in China.

Although mental health has been defined variously by scholars in different cultures, scholars acknowledge that it is more than a lack of mental disorders [98]. Generally, it refers to subjective well-being and a balance among self-satisfaction, independence, capability and competency, achieving potential and coping well with stress and adversity [98]. Several similar terms are used interchangeably in the literature, such as mental disorder, mental illness, mood disturbance, and psychosocial health problems. In short, mental health is "the condition of being free of psychiatric symptoms and having a subjective feeling of well-being" [39], while mental illness refers to "a range of chronic, subjectively unpleasant psychological conditions (listlessness, depression, anxiety, hearing voices not there, feeling one's thoughts being broadcast, and grossly disordered thought) [39].

In terms of the classification of mental disorders, a complete list was given by the ICD-10 classification of mental and behavioral disorders with clinical descriptions and diagnostic guidelines [97]. Based on this classification, several disorders that place a heavy burden on communities are selected by WHO in its 2001 report: "depressive disorders, substance use disorders, schizophrenia, epilepsy, Alzheimer's disease, mental retardation, and disorders of childhood and adolescence" [98]. In terms of the symptoms of mental disorders related to urban environment, the literature review suggests that the most widely discussed symptoms are depression, anxiety, stressfulness, mood state (feeling tense, anxious, worried, afraid and irritable), self-esteem, mental fatigue, aggression and violence [5, 37, 39, 79, 90, 96]. Thus, in suburban neighborhoods, the symptoms that are common in community survey and that could be significantly affected by the urban environment are the focuses, such as depression and anxiety [39].

Moreover, two levels of mental disorders should be clarified: everyday psychopathology (mild or sub-threshold cases) and clinical disorder (serious cases) [39, 93]. The mild symptoms that are not serious enough to be considered as the clinical mental disorders could be termed as risks of mental disorders, such as stress and anxiety. For instance, it is argued that "the evidence linking mental health to stress in daily life is convincing," with higher frequency of stress linked with higher prevalence of mental disorders [2]. So the importance of the establishment of a clinically validated cut-off line should be highlighted in terms of the two levels of mental disorders. However, the clear-cut for dichotomy between cases and non-cases is very difficult for psychiatrists to accept. They have raised two major critiques: (1) the dichotomy classification disregards information about the similarity of some cases and the dissimilarity of others, becoming less sensitive to meaningful changes or differences, (2) separate cut-off points for different groups in the population or more sophisticated techniques are necessary [39].

2.2.1.2 Healthy Neighborhood: Health-Promoting Environment

Conceptually, healthy neighborhood could be interpreted as the neighborhood that promotes the individual and collective health of residents. It has its roots in the concept of supportive environment, which emerged from the First International Conference on Health Promotion Ottawa (1986) and which was elaborated upon at the Sundsvall Conference (1991). Then the concept of "health-promoting environment" has been highlighted in the public health research with the increasing enthusiasm of research on physical activity. With the acknowledgment of the significant effects of the built environment on promoting or discouraging physical activity, public health models have shifted towards emphasizing the health-promoting environment [31]. In the literature, the health-promoting environment is used interchangeably with the supportive environment, indicating a compound of settings, facilities, and programs that encourage physical activity [31]. In addition, Stokols [84] conceptualize a health-promoting environment as "providing environmental resources and interventions that promote enhanced well-being among occupants of an area." Similarly, the environment should provide essential support (such as information and life skills, healthful options among goods and services) in order to help individuals modify or engage in behaviors that maintain their health and reduce their exposure to risk factors [35]. Therefore, a healthy neighborhood could be conceptualized as a health-promoting environment.

Additionally, the term "environment" is clarified for understanding the connotation of healthy neighborhood. It refers to the immediate surroundings of inhabitants, including multiple dimensions in the field of people-environment studies [57], and most scholars agree that the environment has biological, physical, built, social, economic and cultural elements [4, 95]. In the urban context, more emphasis is placed on the physical components such as the natural and built environment in people-environment relationship [75, 91]. In addition, the qualities of place such as population, employment, housing, and transportation, constitute the environment supporting human life in a city [16, 61]. At neighborhood scale, the residential environment is usually the focus of analysis, involving the physical aspect such as housing, streets, green space, and service facilities, as well as the sociocultural aspect such as neighborhood identity and sense of belonging.

In short, a healthy neighborhood is conceptualized as an environmental setting that provides health-promoting and supportive living conditions physically and socially.

2.2.2 *The Human Ecology Perspective for People-Environment Interaction*

For academics or practitioners from both the natural science and human sciences, the ecological perspective provides a conceptual framework for divergent disciplinary concepts and methods and for establishing integrated approaches [55]. Basically, the term "ecology" pertains to the relationships between organisms and their surroundings or environments, which has been interpreted in numerous ways since the late nineteenth century [55, 56, 84, as cited in Hawley 1950]. According to McLaren and Hawe [62, p. 9), the ecological perspective is "a conceptual framework designed to draw attention to individual and environmental determinants of behavior." Using an analogy of treating human habitats as metabolisms, the term, human ecology, was first used by sociologists of the Chicago School of Sociology, attempting to apply the theoretical scheme of plant and animal ecology to the research of human communities [35]. The definitions and interpretations of human ecology have varied considerably during the twentieth century. In terms of the definitions, human ecology is broadly defined as "the interrelations between human beings and their environments, and more particularly on social ecology" [35, p. 270]. Similarly, social ecology is defined as "a framework or set of theoretical principles for understanding the dynamic interrelations among various personal and environmental factors in health" [62, p. 12]. Comparatively, human ecology is closely oriented to the biological processes and the geographic environment, while social ecology pays more attention to the social, institutional, and cultural contexts of people-environment relations [84, 85]. For the application of human ecology, respecting to studies of people-environment relations, various scientific disciplines and professions have taken this perspective, such as archaeology, anthropology, biology, demography, epidemiology, general ecology, geography, law, medicine, political science, psychology and sociology [56, as cited in Bruhn 1974]. In the field of health promotion, it is pointed out that roots of the ecological perspective lie in several disciplines including public health, sociology, psychology, health education and human geography, from where various ecological models generated [35, 74].

For the application of the human ecological perspective to urban studies, it is acknowledged that ecological knowledge should be integrated into urban planning to maintain urban biodiversity and sustainability. For example, many studies have focused on urban ecological systems, especially the urban nature such as greenways, green belts, open space, rivers, and coastlines [49, 65]. The understanding of ecological patterns and processes in urban ecosystems is emphasized, and the importance of interdisciplinary research is pointed out in terms of understanding the complex human influence on urban ecosystems [65]. Another took the perspective of people-environment transactions in the analysis of the relations between behavior and environment, attempting to understand specific factors in the built environment that affect human behaviors [46]. Thus, the ecological perspective could be served as an opportunity for collaboration between the fields of public health and urban planning [46]. Although public health and urban planning address different strategies for health-related issues, with the former focusing on the individual, interpersonal, and multilevel interventions and the latter focusing on measures that shape the built environment, it is recognized that both approaches are fundamental for achieving significant changes in behavior in order to improve human health [46]. In short, the human ecological perspective is an insight for promoting environment and behavior changes through considering individual, interpersonal, community, organizational, policy, and environmental influences, in addition to the dynamic exchange among these factors [46].

The health-promoting environment could be further interpreted from the human ecology perspective in relation to a certain relation between people and environment. Adopting a kind of ecosystem thinking about human settlements, the health-promoting environment should integrate human activities, the physical environment and socio-cultural elements [75]. Both individual and collective well-being are influenced by environments through different paths, indicating that health-promoting environment can be evaluated based on the multiple health outcomes caused by people-environment transactions over a time [84]. Thus, the health-promoting capacity of an environment depends on people-environment transactions, in other words, the level people-environment congruence is an important predictor for health and well-being [84, 85]. As listed in Table 2.1, Stokols [84] summary several environmental qualities that have positive effects on individual and collective well-being and indices for assessing health outcomes at different levels of analysis, offering a preliminary portrait of health-promoting environment. In addition, criteria of health-promoting environment are demonstrated from the three dimensions of health (physical health, mental and emotional well-being, and social cohesion at the organizational and community levels), focusing on environmental resources, behavioral, psychological, and physiological indices (Table 2.1). In this sense, essence of the healthy neighborhood is the congruity between people and environment.

2.2.3 *Human Ecological Frameworks for Understanding Health-Promoting Environments*

Following the discussion on the importance of the human ecological perspective and with the aim of constructing a theoretical foundation for analyzing healthy neighborhood, this section reviews the literature on ecological frameworks and models for understanding the health-promoting environment.

Table 2.1 Dimensions and criteria of health-promoting environment

Facets of healthfulness	Environmental resources	Behavioral, psychological, physiological indices for assessing health outcomes
Physical health	• Injury-resistant design • Ergonomically sound design • Physical comfort, nontoxic and nonpathogenic environment	• Physiologic health • Absence of illness symptoms and injury • Perceived comfort • Genetic and reproductive health
Mental & emotional well-being	• Environmental controllability and predictability • Environmental novelty and challenge • Low distraction • Aesthetic qualities • Symbolic and spiritual elements	• Sense of personal competence, challenge, and fulfillment • Developmental growth • Minimal experience of emotional distress • Strong sense of personal identity and creativity • Feelings of attachment to one's physical and social milieu
Social cohesion at organizational & community levels	• Availability of social support networks • Participatory design and management processes • Organizational flexibility and responsiveness • Economic stability • Low potential for intergroup conflict • Health-promotive media and programming	• High levels of social contact and cooperation • Commitment to and satisfaction with organization and community • Productivity and innovation at organizational and community levels • High levels of perceived quality of life • Prevalence of health-promotive, injury-preventive, environmentally protective behavior

Source Stokols [84, p. 9]

2.2.3.1 The Core Principle: People-Environment Transactions

As discussed, people-environment transactions are crucial for the creation of a health-promoting environment. As an essential principle of the human ecology, the reciprocal determinism has been raised: behavior-environment interactions could mediate the organism's functioning through two ways, first, the environment controls or sets limits on the behavior that occurs in it; second, modifying environmental factors leads to the changes of behavior [35]. This reciprocal idea leads to the credo of health promotion: people should be empowered by given control over the health determinants, in particular, they should be able to adjust their behaviors to the changing environmental conditions, and to modify their environments to the changing behavioral conditions [35]. This reciprocal determinism is authentic since human being ideally take efforts to achieve optimal environment, or to pursue the environment that could fulfill their needs and the accomplish their goals [86].

Accordingly, two types of processes in people-environment transactions are proposed: optimization and passive adaptation processes [88] (Fig. 2.1). In the optimization process, the functional environment could help people to work back on and change it, formulating an ongoing cycle. Passive adaptation process is the dysfunctional and rigid environment that dictates human behavior, and people are unable to work back on and change it (Fig. 2.1). Concerning the function of cognitive representations (e.g., beliefs, attitudes, cognitive maps) and physical and social features of the environment (e.g., material objects, other people, rules), Stokols and Altman [86] see two dimensions of optimization: forms (cognitive and behavioral) and phases (active and reactive) of the transactions. Thus, four modes of people-environment transactions are formulated in Table 2.2: (1) interpretive (active-cognitive): "individual's cognitive representation or construction of the environment"; (2) evaluative (reactive-cognitive): "individual 's evaluation of the situation against predefined standards of quality"; (3) operative (active-behavioral): "individual's movement through or direct impact on the environment"; (4) responsive (reactive-behavioral): "the environment's effects on the individual's behavior and wellbeing" [86, p. 31]. Thus, two assumptions have been raised in understanding the transaction and the optimization: the transactional view emphasizes the effects of environment on people and the reciprocal effects of people on their milieu, the optimization view highlights the endeavors of individuals and groups to create environments that support their prioritized goals, plans and activities maximally [86].

In this sense, health-promoting environment emphasizes the relationship between people and environment towards environmental optimization process. From the human ecological perspective, the health-promoting environment could be interpreted in two ways: the residential environment should support positive health outcomes and eliminate negative ones; the way of formulating the residential environment by people should promote positive health outcomes.

2.2.3.2 Human Ecological Frameworks for Analyzing the Factors Affecting Health

Ecological thinking sees health as a product of the interdependence between the individual and subsystems of the ecosystem, such as family, community, physical and social environment [35]. Since the components of the ecosystem are interdependent, the influential factors for health are multilevel and multispectral [35]. In this case, the theoretical frameworks tease out the key influential factors for health.

First, Urie Brofenbrenner developed a widely applied social ecological framework: the environmental influences on behavior are divided into the micro-, meso-, exo-, and macro-system levels of influence [63]. Based on this, McLeroy et al. [63] develop an ecological framework to understand the individual and environmental

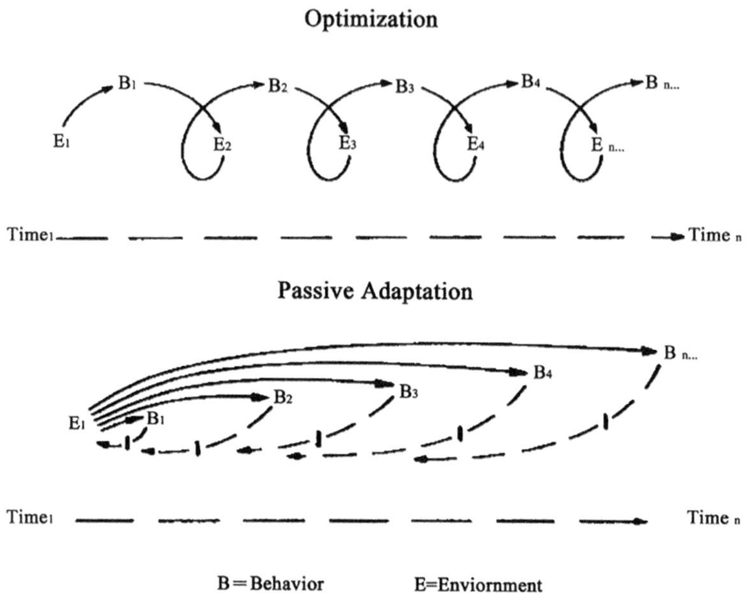

Fig. 2.1 Optimization and passive adaptation process in people-environment transactions [86, p. 31]

Table 2.2 Modes of people-environment transactions and related areas of research

		Form of transaction	
		Cognitive (or symbolic)	Behavioral (or physical)
Phase of transaction	Active	Interpretive Cognitive representation of the spatial environment Personality and the environment	Operative Experimental analysis of economically relevant behavior Human spatial behavior (proxemics)
	Reactive	Evaluative Environmental attitudes Environmental assessment	Responsive Impact of the physical environment Ecological psychology

Source Stokols and Altman [86, p. 32]

determinants of behavior, where behavior is seen as the outcome of interest and is determined by classes of factors:

(1) Intrapersonal factors—characteristics of the individual such as knowledge, attitudes, behavior, self-concept, skills, etc. (2) Interpersonal processes and primary groups—formal and informal social network and social support systems, including the family, work group, and friendship networks. (3) Institutional factors—social institutions with organizational characteristics, and formal (and informal) rules and

regulations for operation. (4) Community factors—relationships among organizations, institutions, and informal networks within defined boundaries. (5) Public policy—local, state, and national laws and policies [63, p. 355].

This ecological model has gained recognition, which helps refocus interests in behavioral determinants of health from intrapersonal determinants of health-related behavior to the social determinants [35].

Second, focusing on people-environment transactions, Stokols [84, 85] propose a socioeconomic analytical framework, focusing on the transactions between individual or collective behavior and health resources or constraints in specific environments [84]. Mindful of the interrelations among environmental conditions and human behavior and well-being, Stokols [84, 85]) point out the following. Environmental conditions have physical, social and cultural dimensions, resulting in a variety of health outcomes, such as physical health status, developmental maturation, emotional well-being, and social cohesion [85, as cited in WHO 1984]. Health is influenced by personal attributes, such as the genetic heritage, psychologic dispositions, and behavioral patterns [85]. The interdependence of environmental conditions[5] and the interconnections between multiply settings and life domains[6] should be both considered in the social ecological analyses [85]. Thus, both personal factors and environmental factors have crucial impacts on health. For the personal factors, three aspects are identified in the interdisciplinary model of environmental and behavioral factors in health: biogenetic, psychological, and behavioral factors [84]. The environmental factors split into three dimensions as well: geographic, architectural and technological, and sociocultural factors [84]. Thus, the analysis of the health-promoting environment should focus on the dynamic interplay between situational and personal factors, and the level of people-environment congruence is a crucial predictor of health and well-being [85]. In addition, two essential facets are highlighted: material resources, which are more directly or objectively observable, such as natural, human-made environmental, economic and technological capital, human resources, which are more abstract and less visible, such as social, moral, and human capital [87].

2.2.3.3 Recaps

Based on the foregoing, several principles could be proposed for studying a health-promoting environment. There is a multilevel and multidimensional concept of health. Figure 2.2 depicts the interactions of the three dimensions of health, and shows that various environmental and personal factors lead to the complexity in conceptualization of the health-promoting environment. A health-promoting environment can be characterized according to the myriad facets of health and to the kinds

---

[5] E.g., the interlinks between physical and social facets of settings, and their independent and joint influences on occupants' well-being.

[6] E.g., independencies existing among immediate and more distance environments.

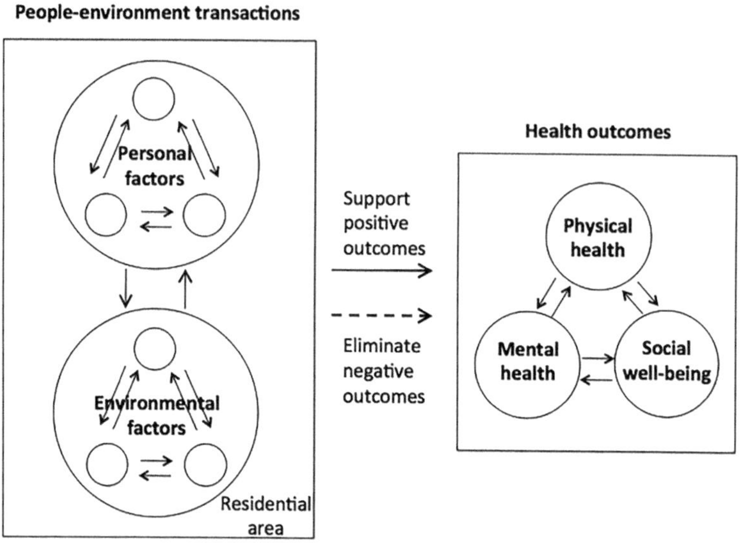

Fig. 2.2 Conceptualization of health-promoting environment

of health outcomes they influence [87]. It is important to examine multiple dimensions of environments and their joint influence on health status at multiple levels, especially the interrelations between the physical and social environment [84].

In addition, the hallmark of the health-promoting environment is people environment congruity, which determine human health and their quality of life. For the personal factors, different component such as biogenetic, psychological, and behavioral factors are related (Fig. 2.2). The various environmental factors are interrelated as well, such as the influences of physical environment on social environment (Fig. 2.2). Between the groups of personal and environmental factors, there could be a network relationship, indicating the interactions between these two groups of factors. These interactions could result in a complex system for assessing the health-promoting environment. (Fig. 2.2) Furthermore, people-environment transactions could produce positive and negative health outcomes, so the cumulative analysis and weighting of both positive and negative features of the living environment is necessary in the assessment of the health-promoting capacity [84] (Fig. 2.2). It is argued that health-promoting capacity of living environment lies on its ability of supporting health benefits those most desirable and important to its residents, and its capacity to eliminate or ameliorate health consequence those most negative and detrimental to individual and social well-being [84].

In addition, a multilevel ecological perspective for studying the health-promoting environment is depicted in Fig. 2.3. The central idea is the reciprocal transactions between personal and environmental factors. The personal factors are similar to the intrapersonal factors proposed by McLeroy et al. [63], including individual choice,

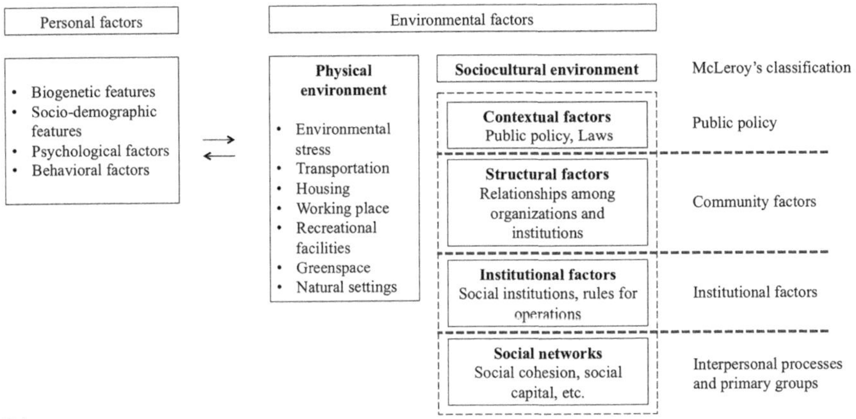

Fig. 2.3 An ecological framework: multiple levels of determinants of health-promoting environment

demographic and socioeconomic status (such as age, gender, education, employment, and household income), biogenetic features, psychological and behavioral factors. There are two dimensions of environmental factors, physical and sociocultural environmental ones. Physical environment includes both the human-made environment and natural environment, including natural settings, ambient environment, housing, working environment, transportation, recreational facilities, and green and open space. The sociocultural environment is comprised of multiple levels as McLeroy et al. [63] proposed: (1) social network at interpersonal level, such as social cohesion, social capital and social support system, (2) institutional factors, such as the rules and regulations for operation; (3) structural (community) factors, focusing on the relations among organizations and institutions, such as power relations; 4) contextual factors, public policy and laws (Fig. 2.3).

Furthermore, the physical environment is determined by the sociocultural environment largely (Fig. 2.3). In the residential neighborhood, physical conditions and interpersonal processes have a direct influence on residents' health. In addition, government regulation and relations among government, developers and residents are significant in the formation of the urban environment. For the effects of the policy, regulatory policies could have a dramatic effect on the health of the population, such as policies containing behavioral incentives and behavioral restrictions [63]. Therefore, in order to understand the health implications of people-environment transactions, it is necessary to explore the related factors at the multiple levels and from multiple dimensions.

2.2.4 Health Determinants at Neighborhood Level

Initially, several issues are clarified. The determinants of health could be considered through two types of mechanisms: selection (drifter) and causation (breeder) [20, 96]. It is important to differentiate the two mechanisms in determining the health determinants, and two hypotheses focusing on causes of mental health are proposed by [17], which is helpful to understanding the selection and causation:

> The selection hypothesis: if the increased prevalence of mental disorders in the new satellite town was caused by selective migration, one would expect the prevalence to stay high, irrespective of possible changes in the social quality of the neighborhood during the follow-up period. [17, p. 531]
>
> The causation hypothesis: if the increased prevalence of mental disorders was caused by unsatisfactory environments, one should expect the prevalence to change parallel with possible changes in the environments. [17, p. 531]

Furthermore, two forms of causation are identified by De Vries et al. [20], which is helpful to comprehend the health effects of living environment: (1) through people's behavior, meaning that living environment could influence human health outcomes via physical activity, such as walking and cycling, (2) through exposure to a specific environment as such, for example, the absence of pollution and pictures of natural settings.

In addition, several shared factors determining health often appear in literature, such as genes, lifestyle and environment. The WHO literature documents the determinant factors of health, including environment, lifestyles, human biology, and health care system (Barton and Tsourou [3], as cited in Lalonde 1974). de Hollander and Staatsen [23] point out that health is determined by the four factors: lifestyle, the physical environment, the social environment, and the endogenous individual attributes (either genetic or acquired during life). Wright et al. [101] categorize four influences of health: behavior (diet, smoking, exercise), environment (housing, education, socioeconomic status, pollution), genes (inherited health potential), and health care (including primary, secondary, and tertiary prevention). While based on a review of health promotion literature [87], posit several "visions" of health determinants: physical layouts of the neighborhood, socioeconomic status, environmental justice level, residents' participation, and health policies and regulations. However, the multi-faceted nature of environmental conditions and health outcomes challenge the work of identifying the most important or pivotal factors that exert the greatest effects on behaviors and health outcomes [87]. Thus, benefited from my educated background of architecture and urban design and focusing on the factors at neighborhood level, the following sections review the health determinants following the human ecological framework, focusing on environmental and personal factors at the levels of intrapersonal factors, interpersonal process and institutional factors. Referring to Fig. 2.3, socio-demographic factors, behavioral factors, physical environmental factors, and sociocultural environmental factors at interpersonal process and at institutional level are the foci.

2.2.5 Environmental Factors

The role of the environment in health and illness has been known since the time of Hippocrates [92], and the link between environment and health has been understood relative comprehensively since the mid-nineteenth century [89]. Among numerous studies on the built environment and health, several themes are discussed in the literature: the built environment and obesity or overweight, the built environment and mental health, the built environment and social capital [73].

Studies of the built environment and obesity often use the body mass index (BMI) to calculate obesity. There is a physical activity related mechanism in the obesity causation, where neighborhood walkability is very important to individual level adjustment [77]. Residents living in the neighborhood that promotes physical activity have a lower BMI [73]. Frank et al. [32] argue that the combination of physical inactivity with other risk behaviors such as improper diet leads to obesity, and the frequency and duration of physical exercise plus dietary pattern are important determinants of overweight. Furthermore, recent studies have emphasized the positive effects of mixed land use, green space and recreational facilities, transit stops, supermarkets, and sport facilities on individual weight status [73, 77].

The longitudinal study conducted by Sarkar et al. [77] shows that retail density and recreational and leisure facilities density are inversely associated with BMI; while the level of land use mix is positively associated with BMI. This controversy could be explained as only the land use mix may not enough to impact BMI [73]. In addition, "the accessibility to green space destination appears no significant association with BMI, and urban design (network and land-use configuration) is more important than open-space and sports facilities in facilitating active ageing" [77, p. 41]. Comprehensively, Papas et al. [67] summarize that the relations between the built environment and obesity depend on the variables of two categories: (1) accessibility and availability of physical activity: population density, whether living on a highway or a street with no sidewalks or not, level of walkability of the neighborhood, sprawl index in county-level, degree of mixed land use, level of greenery, vehicle miles traveled, commuting time, physical activity facilities distribution, (2) accessibility and availability of food sources: the presence of supermarkets and grocery stores, shopping times, numbers of square miles per fast-food restaurant, price of fruits and vegetables.

For the environmental determinants of mental health, Halpern [39] reviewed relative studies published from 1960s to 1980s, and strong associations between the built environment and mental health were revealed in these studies. The built environment could influence mental health through four channels in Halpern's [39] work: first, environmental stress, such as pollution, noise, high social densities, and certain weather conditions. Sullivan and Chang [90] argue that the noisy, crowded and dangerous place could reduce opportunities for social contact, increasing annoyance, distress, anxiety, and in some cases, depression. In addition, the meaning and symbol of the built environment, especially that "the social label process and the discrepancies between individuals' aspirations and their perceived achievements," could have effects on mental health outcomes. It is affirmed that "higher levels of mental illness associated with certain groups and certain building types (notable high-rise) partly may result from discrepancies" [39, p. 206]. The perception of social capital is correlated with mental health [73], and social support and assistance from neighbors could have a positive influence on mental health. Moreover, design characteristics have profound impacts on mental health, and poor neighborhood design is related to distress and depression [18]. Furthremore, public participation in the planning process could also influence the mental health condition of residents. Focusing on the influencing mechanism, the effects of the built environment on mental could be direct and indirect [26]. Physical factors such as "housing, crowding, noise, indoor air quality, and light" could affect mental health directly, while the indirect pathways include "personal control, social supportive relationships and restoration from stress and fatigue" [26, pp. 536–537].

It is admitted that most environmental determinants of health relate to the quality of the living environment and the nature of development [3, 20]. Stemming from the residential environment, Sarkar et al. [78] identify six clusters of principal neighborhood-level determinants of health: density or intensity, diversity or land use mix, destination accessibility, street network morphology and connectivity, access to public transit stops and stations and active transport, and pedestrian-oriented design.

In summary, the following factors of physical environment are discussed in regard to their impact on health: land use morphometric (density and intensity, land use mix), route morphometric (street connectivity or street network morphology, destination accessibility, access to public transit stops, street scale), pedestrian-oriented design, safety, aesthetic qualities, sense of community and social capital [8, 9, 12, 48, 59, 66, 77, 95].

2.2.6 *Personal Factors*

One contextual factor related to individual health outcomes is demographic and socioeconomic status, which could be a context for understanding the effects of environment [92]. The extent to which the environment exerts affects a person's health depends on his or her characteristics, such as age, gender, income, employment, education, and marital status [96]. These characteristics are associated with all-cause mortality and specific diseases, and with individual differences in exposure to negative emotions and the practice of health-related behaviors [Taylor et al. 92, as cited in Adler and Matthews 1994]. The associations between the neighborhood socioeconomic status and health conditions such as BMI, quality of life, chronic conditions and mortality have been affirmed [53], notably, in terms of the mental disorder and demographic and socioeconomic factors, such as the association between unemployment and the incidence of depression [96].

Comprehensively, three dimensions of personal factors that influence health: biogenetic, psychological, and behavioral factors [84]. Since most diseases in twenty-first century are chronic like heart disease, cancer, and diabetes, which could be moderated by the design and construction of living environment, it is accepted that, "in addition to direct hazards from infectious diseases and environmental toxins, human behaviors play a critical role in determining human health" [85]. Thus, due to the involvement of behavioral risk factors in the etiology and progression, the slower-developing chronic illness is known as disease of lifestyle [92].

Health habits or health behaviors are heavily implicated in the onset of illness [92], so person's health-related behaviors are the focus in the behavioral change approach to disease prevention and health promotion [85]. Health-related behaviors include diet, exercise, smoking, alcohol and drug abuse, safe or unsafe sexual practices, personal actions that either decrease or increase the likelihood of bodily injury, compliance with prescribed medical regimens, use of preventive and secondary health service or community services and resources, and health-relevant decisions and actions made on behave of others [84, 85, p. 283,92]. Accordingly, three groups of health-related behaviors are proposed: dietary practices, physical activity, and health screening [66].

Given the focus of this study, it chooses physical activity as the specific health related behaviors for investigation. Notably, individuals living in an active community environment that promotes exercise and activity will be healthier [22]. Physical activity is usually taken as a mediator that being linkages between the built environment and health outcomes, which is both a cause and an effect [32, 74]. Physical activity is defined as "bodily movement produced by skeletal muscles that results in energy expenditure" [68]. There is one related construct should be clarified – physical exercise – a subset of physical activity: planned, structured, and repetitive bodily movement done to improve or maintain one or more components of physical fitness [68].

Focusing on relations between physical activity and health, a growing body of research in public health suggests that participation in regular physical activity can bring substantial and significant health benefits, solve mental health problems and enhance the quality of life for people of all ages [31, 32, 50] According to epidemiologic studies, "physically active adults, as contrasted with their sedentary counterparts, tend to develop and maintain higher levels of physical fitness" [68, p. 403], and there is enormous potential for physical activity to reduce the incidence of chronic disease, overweight and obesity, and premature mortality in the United States [32], and exercise training could improve coronary heart disease risk and other health-related factors [68]. Thus, numerous studies suggest that physical activity is positively associated with both physical and mental health benefits, including reducing mortality rates [69, 79]. Regular physical activity helps maintain normal muscle strength, bone mass, and proper joint structure and function, reduce blood pressure, foster and maintain mental health, such as improve mood and reduce symptoms of depression and anxiety [32, 69]. So the regular physical activity contributes to desirable health outcomes across a variety of physical conditions, better health-related quality of life, better functional capacity and better mood states [69].

In terms of the classification of physical activity, both moderate and vigorous forms of physical activity could generate health benefits, which depend on the three key aspects of physical activity: intensity, duration, and frequency of exercise [31].

The moderate form of physical activity has received more attention to in literature because it is easier for inactive people or the vast majority of the population to begin and maintain over time with little risk of injury [32, 42, 100], and it is beneficial for sedentary individuals most if they increase their activity to the recommended level [68]. Furthermore, for the benchmark of appropriate intensity of physical activity, the Center for Disease Control in US recommends 30 min per day of physical activity for disease prevention. [68] suggest more than 30 min of moderate physical activity on most days of the week to maintain health [31], also note that substantial health gains could be realized if people participate in moderate physical activity regularly, such as 30 min of walking or bicycling, 15 min of jogging or 30–45 min of gardening.

Furthermore, effects of others' behaviors could also affect people's health. For one thing, the relations and interactions between neighbors obviously influence the social network and social cohesion of the neighborhood, contributing to the social wellbeing of the living environment. It is known that "social support and good social relations make an important contribution to health through helping people get emotional and practical resources they need … and feeling cared for, loved, esteemed and valued has powerful protective effect on health" [99]. In this sense, individual health could be the result of the social interactions between neighbors to some extent. For social interactions in neighborhood, two categories could be identified: formal ones (town meetings and purchases at the corner store) and informal or unplanned ones (bumping into a neighbor while running an errand or seeing a "regular" at the coffee shop), which have equal importance in facilitating social capital [24].

2.2.7 *Interactions Between the Built Environment and Physical Activity*

Referring to the importance of reciprocal transactions for health-promoting environment, the impacts of interactions between human behavior and the environment on health outcomes are of great importance from the ecological perspective. As mentioned in Sect. 2.3.1, physical activity is one widely discussed mediator between the built environment and health outcomes. This section focuses on the interactions between the built environment and physical activity.

Consistent with the demonstration on the application of the ecological perspective in the field of public health and urban planning, the relations between human behavior and the environment are the common research interests of two fields. First, the relation between the built environment and human behaviors is a crucial interest for urban planning [42], especially the connection between the built environment and travel behavior (walking or cycling) [76]. The transport theory is conceptually similar to the interpretation of people-environmental transactions, focusing on the factors influencing travel behavior [46]. Urban design theories also "provide useful hypotheses about the elements of the built environment that influence travel behavior, although they do not provide a framework for studying this influence" [46,

p. 15]. Comparatively, public health studies consider the travel behaviors (walking and cycling) as a moderate form of physical activity [32, 76].

Thus, increasing attentions are paid on the effects of environmental change on physical activity, motivating the cooperation between public health and urban planning [46]. However, a divergence between travel behavior and physical activity researchers exists: urban planning emphasizes a utilitarian model, believing that "walking and bicycling derives from the demand for other activities such as work and shopping", while public health prefers a recreational model, noting that "physical activity may be derived from a variety of internal and external factors such as a person's inherent interest in engaging in activity (e.g., jogging, dancing, skiing, team sports) and social support" [46, p. 15]. Thus, referring to the conducting initiatives, two types of physical activities are identified: destination-oriented activity (transportation walking) and leisure-oriented activity (recreation walking), or utilitarian activity and recreational activity [32, 46, 53, 73].

In general, the built environment presents both opportunities for and barriers to participation in physical activity, thereby influencing whether or not people exercise [50]. Scholars usually examine three domains of the built environment: transportation system, land use pattern, urban design characteristics [32, 42]. First, transportation systems (street network, transit systems, specialized networks for non-motorists) determine how well destinations are connected to each other and shape how people move about in the built environment [32]. Transportation environment has impacts on human health through three ways: traffic safety, vehicle pollution, and physical activity, so residents' travel behaviors matter a lot to their health [58]. For the effects of promoting physical activity, the use of walking and bicycling for everyday travel are significant for health and policy concerns, especially in US context [21]. Empirical evidence indicates that regular bicyclists could achieve a healthy level of physical activity through the daily utilitarian travel [21]. Since active transportation (walking, bicycling and public transit) taken by residents is more common in European countries than in the North America and Australia, empirical evidence shows that active transportation could be one of the factors that could explain international differences in the prevalence of obesity [6].

Second, land use pattern represents the arrangement of structures in the built environment and determine the degree of proximity between trip origins and destinations [32], affecting the way of human acting and interacting [50] and further influencing the lifestyle of people [33]. In addition, the combination of transportation system and land use pattern inevitably affect physical activity, such as that the automobile-facilitated environment could reduce physical activity [33], and the connectivity, land use, and residential density are also found to be associated with the likelihood of physical activity as well [100]. Another notable case is urban sprawl and sedentary lifestyle, which is a topic that many scholars make efforts to study. It is believed people living in outlying suburbs get less exercise because of the distances to malls, schools, jobs and residence increases, especially in the US [50].

Third, urban design characteristics could influence how people perceive the built environment and could therefore affect physical activity [32]. Some types of urban design could encourage sedentary living, especially in the design of new residential area in terms of the supposition of driving preference [50]. In contrast, physical activity could be encouraged by proper design of the components of the built environment, such as streets, parks, squares, buildings, lawns, sidewalks, and bus stop shelters. The perception of environment on attractiveness and safety could be determined by the design, which therefore affect people's decision on walking, jogging, cycling or socializing [32]. Specifically, many components of the built environment are found to influence physical activity, such as the presence of absence of sidewalks, heavy traffic, trails, street light, green space, creational facilities, supermarkets, sport facilities, participants' perceived safety, neighborhood upkeep, enjoyable scenery, and architectural features [32, 50, 73, 77]. Additionally, proper combination of the land use, transportation and urban design features could help create active, healthier, and more livable communities [42].

However, the findings on the relations between the built environment and physical activity depend on the different sociocultural contexts. In the US, Frank et al. [32] argue that higher population density and employment could generate more physical activities like walking and biking, and mixed-use development has its greatest relevance at neighborhood scale with respect to physical activity patterns. This is congruent with findings of other Western studies: higher intersection density and higher population density at the destination are associated with a greater likelihood of cycling [100]. The relations between utilitarian walking and density, accessibility, and mix land use are consistent in most previous studies [76]. Studies in developing countries present different findings. In the few studies conducted in China, there is "an inverse relationship between residential density and self-reported recreational physical activity rates among adolescents in Nanjing," but "a positive relationship between residential density and overweight" [19]. In addition, studies conducted in Bogota show that the population density is not conducive to physical activity and that it varies little across the settings observed [19, as cited in Rodriguez et al. 2009, 27]. This incongruence may be because density and its variability differ from place to place and the effects of density on physical activity could vary to geographical regions [100]. Thus, the contradictory findings between developed and developing countries indicate the necessity of further studies on the interaction between the built environment and physical activity with regard to the effects of specific sociocultural characteristics.

2.2.8 Institutional Factors for Creating Health-Promoting Environment

As discussed in Fig. 2.3, a significant sociocultural factor influencing the people-environment congruity is institution. So creating a health-promoting environment for the benefits of human sustainable development resonates with institutions. In Chinese context of suburbanization and the accompanied housing development, this section focuses on the institutional factors that formulate the residential environment in suburbs for middle-income population.

2.2.8.1 Institutional Factors in Suburbanization and Property Development

Contextually, China's contemporary urban development has been driven by both global and local forces, especially the factors of globalization, governance decentralization and marketization [47]. Globally, the paradigm of neoliberalism leads to a political economic transformation in Chinese cities [103]. Combining the features of developmental state with instruments created in the market, Chinese local government is legitimized as a key driver of economic growth [103]. The neoliberalism with Chinese characteristics is then raised [45]. Locally, institutional transitions that generate new process of urban development is the focus of many Chinese urban studies, such as the explanations for the dynamism of entrepreneurial-like government behaviors, including GDP growth mentality, land-based finance, and fiscal policy and incentives [103]. These intuitional reforms result in the decentralization of the state control and marketization of land and housing [47]. Decentralization leads to the power transformation from the state government to local government, which is dedicated to economic growth (Hu [47], as cited in Zhu 2004). Thus, decentralization contributes to a transition from a top down model to a growth regime in the relationship between center cities and suburbs [104]. In addition, land development becomes the basis of suburban growth, which promotes the economic growth and helps to accumulate capital and to relieve financial pressure at the same time [47, 104]. Marketization of land and housing is strategically crucial to urban spatial expansion, resulting in the transitional roles of and interrelations among governments, market and civil society [47]. Thus, institutional mechanisms lead to entrepreneurial-like and neoliberal behaviors of the governments [103], affecting the suburbanization in China greatly.

To understand suburbanization, many scholars use governance as an explanatory tool to investigate the decision making of policies related to suburban expansion and the process of institutionalizing and initiating suburban lifestyles [40]. The effects of the configuration of public and private processes, actors, and institutions on the politics, economics and planning and design of suburbs and daily behaviors are the foci [25]. To identify universal and particular factors determining suburbanization in different urban regions, [25] point out three modalities of institutions for discussing diverse suburbanisms: the state, capital accumulation, private authoritarianism. To be specific, different state forms are determinative in suburbanization; capital has a constitutive role in suburbanization and shaping suburban life, and its governing function could be channeled through the state; the private authoritarian forms of governance is reflected by the gated communities' development [25]. With a careful attention to the relationships among the state, market, and civil society, a coherent logic underlying different forms of suburban space is proposed. land commoditization changes the economic culture, and state entrepreneurialism transforms the role of state, and the power of local elites and local growth coalitions is increased [40, 104].

Promoted by land and housing marketization, diverse spatial forms are formulated, such as informal settlements for rural migrants and formal and master-planned residential enclaves (SCHEs in this study) [103]. The extant literature affirms that the development of SCHEs is associated with neoliberal urbanism, indicating transition of the roles of governments and market [80, 103]. In the property development, the focus of institutional analysis is why and how government intervene the property market [52]. For the changing roles of state, there are two distinctive stages by 1998 in relation to the dynamic land marketization. The state decentralized its power to local governments in 1980s to promote the marketization of land use rights, while the state enhanced its regulation since the late 1990s by tightening land supply and building a transparent market for land use right transfers [47]. In the leasing of land for property development, developers provide the revenue sources for governments by purchasing the land use rights [47].

In housing development, [7] housing process model provides a framework for explaining the roles of different stakeholders. The framework classifies the housing process into six periods: inception, design and construction, building use, refurbishment and demolition, which have significant influences on the living environment [13]. Scrutinizing the roles of and the relationships among different stakeholders in these stages could help to identify the institutional factors formulating the residential environment. For example, with China's housing marketization, the state transfers its role from a housing provider to a policy maker, establishing a housing market and promoting the provision of commodity housing [47, 108]. Developers use gated communities as a place-branding tactic to increase the attractiveness of suburban housing, while local government endorses the SCHEs development for increasing land revenue and capital mobilization to fund the infrastructure in suburban development [103]. In addition, the forces of market demands and consumer behaviors have driven the housing development [47], indicating the influential roles of developers and consumers. Furthermore, planners provide their professional techniques to achieve the development goals of the developers under the regulations of governments. In this sense, suburban housing development is both land-driven and consumer-driven, and local government, developers, and planning professionals play important roles in place creation. Moreover, in housing management, the self-governance is only confined to the role of property management, with the dominance of residents' committee in management affairs and the underdeveloped civil society [103]. To summarize, in general, the role of the state is pivotal in terms of the co-engineered with other actors, developer is a significant force in shaping suburban landscape; rising consumerism leads to the emergence and the necessity of branding, packaging, imagery, and marketing [104]. In short, the roles of and the state, the market and the consumers are of strategic importance to the housing development in terms of the formation of residential environment and the way of life of the middle-income people.

2.2.8.2 Institutional Factors in Health Promotion

For public health services, such as the population-based efforts to investigate community health threats, promotion of healthy lifestyles, prevention of disease and injury, and ensuring the quality of the living environment for good health, institutional factors are significant in terms of their performance [60]. The structural and organizational characteristics of the public health system mainly include the size of the system and the type of government authority and control within the system. Specifically, large public health systems have advantages in the performance of the system in terms of organizations that could participate in public health activities, in addition, the decentralized or centralized organizational structures are influential for the system performance [60]. Urban health depends on institutional mechanisms through which societies manage the courses of event [10, 11]. Discussions on institutional mechanisms in health domain have been increasing, especially the governance for health. Since health depends on people's everyday lives, a new perspective has been raised in the governance of health in relation to the creation of health living environment [54]. Stemmed from the concept of "stewardship" as framed by WHO in 2000, governance is used to understand the institutional determinants of health [10]. It has been pointed out that most health determinants are found in non-health sectors, and health and non-health sectors, public and private sectors are involved in the common interest of health [54]. In addition, the good governance for health sees health as a human right, an element of well-being, a global public good, and a matter of social justice and equity, guiding the policy-making for health [54, 112].

With the aim of embedding health in policies of urban development, institutional factors in urban development are crucial. However, there are inevitable institutional barriers of integrating health and urban development policies, including "the uniformity or autonomy of departments, connectivity of departments (different level of interdependent of departments), formality of relationships (formal or informal networks), and instruments (the use of planning, formal regulations and contracts)" [83]. It is argued that policy integration should be promoted in terms of synergies between sectors, consistency between policies in different sectors and policy-making process, especially in relation to transportation, health and environmental issues [83]. In addition, institutional innovation in the governance structures is necessary to enable the collaboration between health and non-health sectors, and between public and private sectors.

2.3 Conclusion

This chapter develops a comprehensive theoretical foundation for understanding suburbanization, healthy neighborhoods and health determinants in neighborhood environment from the lens of human ecology. For comprehending suburbs, China's suburbanization is reviewed, which is unique in terms of the initiatives of urban expansion, physical characteristics of the built environment and the socioeconomic status of neighborhoods. China's suburbs are featured by greater governments' interventions, high density and transportation-oriented development, and demographic diversity. In addition, residential suburbanization is accompanied with the boom of SCHEs, which change both urban landscape and the lifestyles of the new middle-class.

From the human ecological perspective, health is seen as an outcome of people-environment transactions at multiple levels and dimensions, physically and mentally. A healthy neighborhood is conceptualized as a health-promoting environment, and its essence is people-environment congruity, which is important to the environmental optimization process. Literature review suggests a multilevel and multidimensional ecological framework for understanding the health-promoting environment in relation to personal and environmental factors. In particular, socio-demographic factors, behavioral factors, physical environmental factors, and sociocultural environmental factors at interpersonal process and at institutional level are the foci of this study.

According to the human ecological model of healthy neighborhoods, health determinants at the neighborhood level are identified from the literature. The environmental factors that are associated with health involve: transportation system, land use pattern, residential amenities, and design features. The personal factors affecting health include socio-demographic characteristics, daily routine, the physical activity, and social interactions. Studies of the interaction between people and environment pay much attention to the effects of the built environment on physical activity, while there is no consensus in the results between developed and developing countries. In addition, institutional factors affecting the creation of health-promoting environment are reviewed in relation to China's suburbanization and housing development. Institutional transitions led by neoliberalism, governance decentralization and marketization change the roles of and interactions among the governments, the developers and the civil society to form a new institutional environment, affecting physical and sociocultural characteristics of the residential environment and the resultant way of life, and in turn, residents' health.

Although a rich body of literature focuses on the determinants of health from either environmental or personal perspectives, few studies use the human ecological perspective to explain the health implications of people-environment transactions. Most studies reveal the effects of the environment on behavior rather than the inverse. It is important to examine the two ways relationships. In addition, fewer studies have identified institutional factors shaping the health determinants in suburbanization and housing development, and limited concerns in literature focus on the demand of middle-class in housing marketization, from the lens of health-promoting environment.

References

1. Adell G (1999) Theories and models of the peri-peri urban interface: a changing conceptual landscape. https://discovery.ucl.ac.uk/id/eprint/43/1/DPU_PUI_Adell_THEORIES_MODELS.pdf
2. Bancila D, Singhammer J (2013) The variation of the stress–mental health relationship by ethnicity and gender. Int J Ment Health Promot 15(3):178–192. https://doi.org/10.1080/14623730.2013.836833
3. Barton H, Tsourou C (2000) Healthy urban planning: a WHO guide to planning for people. Taylor & Francis
4. Barton H, Tsourou C (2013) Healthy urban planning. Routledge
5. Barton J, Pretty J (2010) What is the best dose of nature and green exercise for improving mental health? a multi-study analysis. Environ Sci Technol 44(10):3947–3955. https://doi.org/10.1021/es903183r
6. Bassett DR, Pucher J, Buehler R, Thompson DL, Crouter SE (2008) Walking, cycling, and obesity rates in Europe, North America, and Australia. J Phys Act Health 5(6):795–814. https://doi.org/10.1123/jpah.5.6.795
7. Bhatti M (1994) Environmental futures and the housing question. In: Bhatti M, Brooke J, Gibson M (eds) Housing and the environment: a new agenda. Chartered Institute of Housing, Coventry, pp 14–33
8. Bonaiuto M, Fornara F, Bonnes M (2003) Indexes of perceived residential environment quality and neighbourhood attachment in urban environments: a confirmation study on the city of Rome. Landsc Urban Plan 65(1):41–52. https://doi.org/10.1016/S0169-2046(02)00236-0
9. Bonaiuto M, Fornara F, Bonnes M (2006) Perceived residential environment quality in middle- and low-extension Italian cities. Eur Rev Appl Psychol 56(1):23–34. https://doi.org/10.1016/j.erap.2005.02.011
10. Burris S, Hancock T, Lin V, Herzog A (2007) Emerging strategies for healthy urban governance. J Urban Health 84(1):154–163. https://doi.org/10.1007/s11524-007-9174-6
11. Burris S, Ompad D (2010) Healthy urban governance. Urban Heal: Global Perspect 18:355
12. Carp FM, Zawadski RT, Shokrkon H (1976) Dimensions of urban environmental quality. Environ Behav 8(2):239–264. https://doi.org/10.1177/001391657682004
13. Chiu RLH (2000) Environmental sustainability of Hong Kong's housing system and the housing process model. Int Plan Stud 5(1):45–64. https://doi.org/10.1080/135634700111819
14. Chung CJ, Inaba J, Koolhaas R, Leong ST (2001) Great leap forward: Harvard Design School project on the City. Taschen GmbH
15. Clapson M, Hutchison R (2010) Introduction: suburbanization in global society. In: Clapson M, Hutchison R (eds), Suburbanization in global society, Vol 10. Emerald Group Publishing Limited, pp 1–14. https://doi.org/10.1108/S1047-0042(2010)0000010003
16. Cowan R, Rogers L (2005) The dictionary of urbanism, Vol 67. Streetwise Press Tisbury
17. Dalgard OS, Tambs K (1997) Urban environment and mental health: a longitudinal study. Br J Psychiatry 171(6):530–536. https://doi.org/10.1192/bjp.171.6.530
18. Dannenberg AL, Frumkin H, Jackson RJ (2011) Making healthy places: designing and building for health, well-being, and sustainability. Island Press
19. Day K, Alfonzo M, Chen Y, Guo Z, Lee KK (2013) Overweight, obesity, and inactivity and urban design in rapidly growing Chinese cities. Health Place 21:29–38. https://doi.org/10.1016/j.healthplace.2012.12.009
20. De Vries S, Verheij RA, Groenewegen PP, Spreeuwenberg P (2003) Natural environments—healthy environments? an exploratory analysis of the relationship between greenspace and health. Environ Plan A 35(10):1717–1731
21. Dill J (2009) Bicycling for transportation and health: the role of infrastructure. J Public Health Policy 30(1):S95–S110. https://doi.org/10.1057/jphp.2008.56
22. Doyle S, Kelly-Schwartz A, Schlossberg M, Stockard J (2006) Active community environments and health: the relationship of walkable and safe communities to individual health. J Am Plann Assoc 72(1):19–31. https://doi.org/10.1080/01944360608976721

23. de Hollander AE, Staatsen BA (2003) Health, environment and quality of life: an epidemiological perspective on urban development. Landsc Urban Plan 65(1–2):53–62
24. Eicher C, Kawachi I (2011) Social capital and community design. In: Dannenberg AL, Frumkin H, Jackson RJ (eds), Making healthy places: designing and building for health, well-being, and sustainability. Island Press/Center for Resource Economics, pp. 117–128. https://doi.org/10.5822/978-1-61091-036-1_8
25. Ekers M, Hamel P, Keil R (2012) Governing suburbia: modalities and mechanisms of suburban governance. Reg Stud 46(3):405–422. https://doi.org/10.1080/00343404.2012.658036
26. Evans GW (2003) The built environment and mental health. J Urban Health 80(4):536–555. https://doi.org/10.1093/jurban/jtg063
27. Ewing R, Cervero R (2010) Travel and the built environment: a meta-analysis. J Am Plann Assoc 76(3):265–294. https://doi.org/10.1080/01944361003766766
28. Feng J (2001) The progress and prospect of the research on suburbanization in China. Hum Geogr 16(6):30–35
29. Feng J, Zhou Y, Wu F (2008) New trends of suburbanization in Beijing since 1990: from government-led to market-oriented. Reg Stud 42(1):83–99. https://doi.org/10.1080/00343400701654160
30. Forsyth A (2012) Defining suburbs. J Plan Lit 27(3):270–281. https://doi.org/10.1177/0885412212448101
31. Frank L, Engelke PO (2001) The built environment and human activity patterns: exploring the impacts of urban form on public health. J Plan Lit 16(2):202–218. https://doi.org/10.1177/08854120122093339
32. Frank, L., Engelke, P., & Schmid, T. (2003). *Health and Community Design: The Impact Of The Built Environment On Physical Activity*. Island Press.
33. Frumkin H, Frank LD, Jackson RJ (2004) Urban sprawl and public health: designing, planning, and building for healthy communities. Island Press
34. Gotham KF (2013) International perspectives on suburbanization: a post-suburban world? Contemp Sociol 42(2):268–269. https://doi.org/10.1177/0094306113477381kk
35. Green LW, Richard L, Potvin L (1996) Ecological foundations of health promotion. Am J Health Promot 10(4):270–281. https://doi.org/10.4278/0890-1171-10.4.270
36. Gregory D, Johnston R, Pratt G, Watts M, Whatmore S (2009) The dictionary of human geography. Wiley-Blackwell
37. Groenewegen PP, van den Berg AE, Maas J, Verheij RA, de Vries S (2012) Is a green residential environment better for health? if so, why? Ann Assoc Am Geogr 102(5):996–1003. https://doi.org/10.1080/00045608.2012.674899
38. Gu C (1995) Urban preferiy of China's big cities (Zhong guo da cheng shi bian yuan qu yan jiu). Science Press
39. Halpern D (1995) Mental health and the built environment: more than bricks and mortar? Taylor & Francis US
40. Hamel P, Keil R (2015) Suburban governance: a global view. University of Toronto Press
41. Hancock T (1997) Healthy cities and communities: past, present, and future. Natl Civ Rev 86(1):11–21
42. Handy SL, Boarnet MG, Ewing R, Killingsworth RE (2002) How the built environment affects physical activity: views from urban planning. Am J Prev Med 23(2):64–73. https://doi.org/10.1016/S0749-3797(02)00475-0
43. Harris R (2010) Meaningful types in a world of suburbs. In: Clapson M, Hutchison R (eds) Suburbanization in global society, Vol 10. Emerald Group Publishing Limited, pp 15–47. https://doi.org/10.1108/S1047-0042(2010)0000010004
44. Harris R, Larkham P (2003) Changing suburbs: foundation, form and function. Routledge, Abingdon
45. Harvey D (2007) A brief history of neoliberalism. Oxford University Press, USA
46. Hoehner CM, Brennan LK, Brownson RC, Handy SL, Killingsworth R (2003) Opportunities for integrating public health and urban planning approaches to promote active community environments. Am J Health Promot 18(1):14–20. https://doi.org/10.4278/0890-1171-18.1.14

47. Hu R (2013) Drivers of China's urbanisation and property development. Australasian J Reg Stud 19(2):156–180. https://doi.org/10.3316/informit.579804722656285
48. Hu W (2007) Residential area planning principle and design method. China Architecture & Building Press
49. Ignatieva M, Stewart GH, Meurk C (2011) Planning and design of ecological networks in urban areas. Landscape Ecol Eng 7(1):17–25. https://doi.org/10.1007/s11355-010-0143-y
50. Jackson RJ, Kochtitzky C (2003) Creating a healthy environment: the impact of the built environment on public health. Sprawl Watch Clearinghouse Monograph Series, 19
51. Jackson RJ, Stacy S (2011) Designing healthy communities. John Wiley & Sons
52. Kauko T (2012) An institutional analysis of property development, good governance and urban sustainability. Eur Plan Stud 20(12):2053–2071. https://doi.org/10.1080/09654313.2012.722926
53. Kerr J, Rosenberg D, Frank L (2012) The role of the built environment in healthy aging: community design, physical activity, and health among older adults. J Plan Lit 27(1):43–60. https://doi.org/10.1177/0885412211415283
54. Kickbusch I, Gleicher DE (eds) (2012) Governance for health in the 21st century. WHO Regional Office for Europe
55. Lawrence RJ (2002) Healthy residential environments. In: Handbook of environmental psychology, pp 394–412
56. Lawrence RJ (2003) Human ecology and its applications. Landsc Urban Plan 65(1):31–40. https://doi.org/10.1016/S0169-2046(02)00235-9
57. Lawrence RJ (2012) People-environment studies. In: The SAGE handbook of housing studies, pp 230–244
58. Litman T (2003) Integrating public health objectives in transportation decision-making. Am J Health Promot 18(1):103–108. https://doi.org/10.4278/0890-1171-18.1.103
59. Marans RW (2003) Understanding environmental quality through quality of life studies: the 2001 DAS and its use of subjective and objective indicators. Landsc Urban Plan 65(1):73–83. https://doi.org/10.1016/S0169-2046(02)00239-6
60. Mays GP, McHugh MC, Shim K, Perry N, Lenaway D, Halverson PK, Moonesinghe R (2006) Institutional and economic determinants of public health system performance. Am J Public Health 96(3):523–531. https://doi.org/10.2105/AJPH.2005.064253
61. McClure WR, Bartuska TJ (2007) The built environment: a collaborative inquiry into design and planning. John Wiley & Sons
62. McLaren L, Hawe P (2005) Ecological perspectives in health research. J Epidemiol Community Health 59(1):6–14. https://doi.org/10.1136/jech.2003.018044
63. McLeroy KR, Bibeau D, Steckler A, Glanz K (1988) An ecological perspective on health promotion programs. Health Educ Q 15(4):351–377. https://doi.org/10.1177/109019818801500401
64. Moser G (2009) Quality of life and sustainability: toward person–environment congruity. J Environ Psychol 29(3):351–357. https://doi.org/10.1016/j.jenvp.2009.02.002
65. Niemelä J (1999) Ecology and urban planning. Biodivers Conserv 8(1):119–131. https://doi.org/10.1023/A:1008817325994
66. Northridge ME, Sclar ED, Biswas P (2003) Sorting out the connections between the built environment and health: a conceptual framework for navigating pathways and planning healthy cities. J Urban Health 80(4):556–568. https://doi.org/10.1093/jurban/jtg064
67. Papas MA, Alberg AJ, Ewing R, Helzlsouer KJ, Gary TL, Klassen AC (2007) The built environment and obesity. Epidemiol Rev 29(1):129–143. https://doi.org/10.1093/epirev/mxm009
68. Pate RR, Pratt M, Blair SN, Haskell WL, Macera CA, Bouchard C, Buchner D, Ettinger W, Heath GW, King AC, Kriska A, Leon AS, Marcus BH, Morris J, Paffenbarger RS Jr, Patrick K, Pollock ML, Rippe JM, Sallis J, Wilmore JH (1995) Physical activity and public health: a recommendation from the centers for disease control and prevention and the American College of Sports Medicine. JAMA 273(5):402–407. https://doi.org/10.1001/jama.1995.03520290054029

69. Penedo FJ, Dahn JR (2005) Exercise and well-being: a review of mental and physical health benefits associated with physical activity. Curr Opin Psychiatry 18(2):189
70. Phelps NA, Wu F (2011) International perspectives on suburbanization. Springer
71. Pryor RJ (1968) Defining the rural-urban fringe. Soc Forces 47(2):202–215
72. Ren X (2010) Territorial expansion and state rescaling: A critique of suburbanization studies in China. In: Clapson M, Hutchison R (eds), Suburbanization in global society, Vol 10. Emerald Group Publishing Limited, pp. 347–366. https://doi.org/10.1108/S1047-0042(2010)0000010017
73. Renalds A, Smith TH, Hale PJ (2010) A systematic review of built environment and health. Fam Community Health 33(1):68–78. https://doi.org/10.1097/FCH.0b013e3181c4e2e5
74. Richard L, Gauvin L, Raine K (2011) Ecological models revisited: their uses and evolution in health promotion over two decades. Annu Rev Public Health 32(1):307–326. https://doi.org/10.1146/annurev-publhealth-031210-101141
75. Roseland M (2000) Sustainable community development: integrating environmental, economic, and social objectives. Prog Plan 54(2):73–132. https://doi.org/10.1016/S0305-9006(00)00003-9
76. Saelens BE, Handy SL (2008) Built environment correlates of walking: a review. Med Sci Sports Exerc 40(7):S550. https://doi.org/10.1249/MSS.0b013e31817c67a4
77. Sarkar C, Gallacher J, Webster C (2013) Built environment configuration and change in body mass index: the Caerphilly Prospective Study (CaPS). Health Place 19:33–44. https://doi.org/10.1016/j.healthplace.2012.10.001
78. Sarkar C, Webster C, Gallacher J (2014) Healthy cities: public health through urban planning. Edward Elgar Publishing
79. Scully D, Kremer J, Meade MM, Graham R, Dudgeon K (1998) Physical exercise and psychological well being: a critical review. Br J Sports Med 32(2):111–120. https://doi.org/10.1136/bjsm.32.2.111
80. Shen J, Wu F (2012) The Development of master-planned communities in Chinese suburbs: a case study of Shanghai's Thames town. Urban Geogr 33(2):183–203. https://doi.org/10.2747/0272-3638.33.2.183
81. Shen J, Wu F (2013) Moving to the suburbs: demand-side driving forces of suburban growth in China. Environ Plann A: Econ Space 45(8):1823–1844. https://doi.org/10.1068/a45565
82. Smith T, Nelischer M, Perkins N (1997) Quality of an urban community: a framework for understanding the relationship between quality and physical form. Landsc Urban Plan 39(2):229–241. https://doi.org/10.1016/S0169-2046(97)00055-8
83. Stead D (2008) Institutional aspects of integrating transport, environment and health policies. Transp Policy 15(3):139–148. https://doi.org/10.1016/j.tranpol.2007.12.001
84. Stokols D (1992) Establishing and maintaining healthy environments: toward a social ecology of health promotion. Am Psychol 47:6–22. https://doi.org/10.1037/0003-066X.47.1.6
85. Stokols D (1996) Translating social ecological theory into guidelines for community health promotion. Am J Health Promot 10(4):282–298. https://doi.org/10.4278/0890-1171-10.4.282
86. Stokols D, Altman I (1987) Environmental psychology. Percep Cognit Image City 12:23–90
87. Stokols D, Grzywacz JG, McMahan S, Phillips K (2003) Increasing the health promotive capacity of human environments. Am J Health Promot 18(1):4–13. https://doi.org/10.4278/0890-1171-18.1.4
88. Stokols D, Lejano RP, Hipp J (2013) Enhancing the resilience of human-environment systems: a social ecological perspective. Ecol Soc 18(1):7. https://doi.org/10.5751/ES-05301-180107
89. Stone D (2006) Sustainable development: convergence of public health and natural environment agendas, nationally and locally. Public Heal 120(12):1110–1113. https://doi.org/10.1016/j.puhe.2006.10.011
90. Sullivan WC, Chang CY (2011) Mental health and the built environment. In: Dannenberg AL, Frumkin H, Jackson RJ (eds) Making healthy places: designing and building for health, well-being, and sustainability. Island Press/Center for Resource Economics, pp 106–116. https://doi.org/10.5822/978-1-61091-036-1_7

91. Sutton P (2004) A perspective on environmental sustainability. Paper on the Victorian Commissioner for Environmental Sustainability 1:32
92. Taylor SE, Repetti RL, Seeman T (1997) Health psychology: what is an unhealthy environment and how does it get under the skin? Annu Rev Psychol 48(1):411–447. https://doi.org/10.1146/annurev.psych.48.1.411
93. The WHO World Mental Health Survey Consortium (2004) Prevalence, severity, and unmet need for treatment of mental disorders in the World Health Organization world mental health surveys. JAMA 291(21):2581–2590. https://doi.org/10.1001/jama.291.21.2581
94. Thompson FML (1982) The rise of suburbia, Vol 3. Burns & Oates
95. van Kamp I, Leidelmeijer K, Marsman G, de Hollander A (2003) Urban environmental quality and human well-being: towards a conceptual framework and demarcation of concepts; a literature study. Landsc Urban Plan 65(1):5–18. https://doi.org/10.1016/S0169 2046(02)00232-3
96. Verheij RA (1996) Explaining urban-rural variations in health: a review of interactions between individual and environment. Soc Sci Med 42(6):923–935. https://doi.org/10.1016/0277-9536(95)00190-5
97. WHO (1992) The ICD-10 classification of mental and behavioural disorders: Clinical descriptions and diagnostic guidelines. World Health Organization. https://apps.who.int/iris/handle/10665/37958
98. WHO (2001) The World Health Report 2001: mental health: new understanding, new hope
99. Wilkinson RG, Marmot M (2003) Social determinants of health: the solid facts. World Health Organization
100. Winters M, Brauer M, Setton EM, Teschke K (2010) Built environment influences on healthy transportation choices: bicycling versus driving. J Urban Health 87(6):969–993. https://doi.org/10.1007/s11524-010-9509-6
101. Wright J, Williams R, Wilkinson JR (1998) Development and importance of health needs assessment. BMJ 316(7140):1310–1313. https://doi.org/10.1136/bmj.316.7140.1310
102. Wu F (2012) Suburban Beijing: housing and consumption in contemporary China, by Friederike Fleischer. China J 67:238–239. https://doi.org/10.1086/665777
103. Wu F (2016) Housing in Chinese urban villages: the dwellers, conditions and tenancy informality. Hous Stud 31(7):852–870. https://doi.org/10.1080/02673037.2016.1150429
104. Wu F, Shen J (2015) Suburban development and governance in China. In: Suburban governance: a global view. University of Toronto Press
105. Xie S, Ning Y (2002) Urbanization and suburbanization: the dual engines to spatial change of a Chinese metropolis during the transitional era: a case study of Guangzhou. City Planning Review 27(11):24–38
106. Yadav CS (1987) Rural-urban fringe, Vol 9. Concept Publishing Company
107. Yeung CW (1984) Large-scale private housing development in the suburban area in Hong Kong a case study of Fairview Park. University of Hong Kong, Centre of Urban Studies and Urban, Planning
108. Zhang X (2002) Governing housing in China: state, market and work units. J Housing Built Environ 17(1):7–20
109. Zhang X, Shen J, Gao X (2021) Towards a comprehensive understanding of intercity cooperation in China's city-regionalization: a comparative study of Shenzhen-Hong Kong and Guangzhou-Foshan city groups. Land Use Policy 103:105339
110. Zhang X, Li Y, Zhou C, Luan X, Yuan F (2023a) Rescaling of the land regime in the making of city-regions: a case study of China's Pearl River delta. Urban Stud 60(3):483–500
111. Zhang X, Sun Y, Luan X (2023b) Regulation flexibility and legitimacy building in governing intercity railways: the polymorphous role of the Chinese provincial government. J Urban Aff, 1–16. https://doi.org/10.1080/07352166.2022.2141641
112. Zhang X, Zhou Z, Yang Y (2023c) Stepwise transition of urban entrepreneuralism in cross-boundary statecraft: a qualitative study of an enclave industrial park in Zhejiang Province, China. Trans Plan Urban Res 2(1):85–102

113. Zhou C, Xu X (1996) A study of the spatial characteristics of population change in Guangzhou City. Econ Geogr 16(2):25–30
114. Zhou Y, Logan JR (2008) Growth on the edge: the new Chinese metropolis. In: Urban China in transition. John Wiley & Sons, Ltd., pp 140–160. https://doi.org/10.1002/9780470712870.ch6

Chapter 3
Health Issues of Suburban Living in Guangzhou (China): An Empirical Review

To understand the suburban living and its health effects in China, the housing development that is companied with suburbanization should be paid attention to, which is important to the creation of health-promoting environment. Although the quality of commodity housing has improved greatly after more than three decades' development, little attention has been paid to the health implications of living in commodity housing estates, especially in suburban areas. Market-led suburbanization may have neglected the health of residents during urban expansion; moreover, the suburban lifestyle shaped by suburbanization and suburban housing planning and design may have various impacts on human health. Thus, it is necessary to know what are the empirical issues in housing development in terms of promoting health benefits of its residents, and what are the most popular health issues in urban residents of China.

This chapter is comprised of two sections, reviewing the housing development issues that are adverse to healthy neighborhoods and the resultant health tolls. In the first section, the characteristics of commodity housing development in China are summarized, based on which SCHEs are proposed as the focus of this study. Afterwards, the development of SCHEs in Guangzhou is elaborated, contributing to the discussion on the development issues adverse to the healthy neighborhoods. Particular efforts are made to demonstrate the institutional changes and population attributes in SCHEs development in Guangzhou. In the second section, wildly discussed disadvantages in suburban living are pointed out, followed by which the health tolls of urban environment and suburban living in China are reviewed, focusing on two popular illness of individual health: overweight and obesity, and mental disorders. This chapter focuses on the housing development issues that are adverse to a healthy neighborhood and the health issues of suburban living. Although the quality of commodity housing has improved greatly after more than three decades' development, little attention has been paid to the health implications of living in commodity housing estates, especially in suburban areas. Market-led suburbanization may have neglected the health of residents during urban expansion; moreover, the pattern of suburban living formulated by suburbanization could threaten the human health.

T. Zhang, *Towards Healthy Settlements*, Urban Sustainability,
https://doi.org/10.1007/978-981-97-1207-6_3

3.1 Issues in SCHEs Development Against Healthy Neighborhood

3.1.1 Commodity Housing Development in China: A Trend to Suburban Living

As a major outcome of urbanization and urban decentralization, commodity housing development is crucial to the creation of healthy neighborhoods. Reviews of its origin and development background could help to capture the nature of SCHEs, contributing to the identification of the housing development issues that are adverse to the creation of a health-promoting environment.

3.1.1.1 Context of the Commodity Housing Development

Most studies of commodity housing were conducted around 2000, addressing housing reform and housing commodification [8, 40, 59, 68, 73, 76]. There are four stages of commodity housing development: (a) 1979–1984: the beginning stage of the economic reform; (b) 1985–1991: the planned commodity economy; (c) 1992–2000: the early period of the socialist market economy [40]; (d) after 2001: the improved socialist market economy and housing market boom. During 1979–1984, motivated by the economic reform that gave market forces and private enterprise a larger role in housing production and consumption, the real estate industry began to develop [59].

In the second period (1985–1991), housing commodification accelerated though the administrative allocation system (officially called welfare distribution system) largely remained intact [40]. Led by Shenzhen, commodity housing estates developed increasingly, leading to improvements in both housing quantity and housing quality. While the major purchasers of commodity housing estates were institutional groups, who allocated the purchased housing to their workers and staff as a type of in-kind welfare [40]. In terms of development mode, owing to the problems caused by the conventional enterprise- based or institution-based housing development, there was a transformation moving towards a comprehensive development approach. It means that the whole housing lifecycle and unified management of housing development, ranging from surveying, planning and design to land allocation, demolition, site preparation and infrastructure construction[1] to maximize the economic, social and environmental benefits [40, 59]. The real estate industry had adopted the comprehensive housing development approach by the mid-1980s. However, as real estate enterprises are profit-oriented, commodity housing development was influenced by the intention to maximizing profits leading to, e.g., over pursuit of plot ratio, going

[1] Such as road construction, water supply, sewage, power supply, gas supply, heating and communications.

against the original purpose of the comprehensive development in term of ensuring economic, social and environmental sustainability [40].

In the third stage of 1992–2000, following the economic improvement and rectification as well as accelerated urban development during 1988–1991, the real estate industry boomed [40, 59]. Official statistics show that the percentages of commodity housing investment in total urban housing investment increased from 27 to 60% from 1991 to 1994 [59]. Influenced by the unprecedented growth of the macro economy and the incomplete market mechanism, purchasing commodity housing became a way of speculating, and housing price increase resulting in lots of problems such as the oversupply of land, exceeding the capability of the construction sector to absorb, and an oversupply of luxurious apartments and villas. In this context, the government issued a series of macro-regulatory policies from 1993 to promote housing reform more aggressively. However, to help overcome the economic damage inflicted by the Asian Financial Crisis of 1997, the real estate industry was designated by the government as a new economic sector to resuscitate economic growth. By 2000, the proportion of commodity housing in the whole housing stock had increased to 67.7% [75], and real estate markets had significantly pushed urban development forward ([75]).

From surveying, planning and design to land allocation, demolition, site preparation and infrastructure construction to maximize its economic, social and environmental benefits [40, 59]. The real estate industry had adopted comprehensive housing development h by the mid-1980s. However, as real estate enterprises are profit-oriented, commodity housing development was influenced by the impetus of maximum profits such as the over pursuit of the plot ratio, going against the original purpose of the comprehensive development [40].

Accordingly, a new housing system was established with the aims of conforming to the marketizing socialist economy, attempting to complete the process of housing commodification and widen the bases of producing consumption financing to multi-stakeholders, improving living conditions and satisfying the housing demands of urban residents [40]. As a result, the housing investment mechanisms contribute to the establishment of Housing Provident Fund, monetization of housing distribution, and establishment of multi-layered social security housing system[2] [40], the transition from housing welfare allocation to housing commoditization was officially completed between 1998 and 2000 [6, 40]. Under the new housing system, commodity housing estates became the only choice for the middle-upper income households and above, as well as for families not eligible to join the social security housing system, nor for the allocation of the meager social housing. For these reasons, different types of commodity housing were developed to satisfy the range of demands in the form of high-rise housing, multi-story housing, low-rise high-density housing, terrace houses and villas.

[2] A new concept of "basic housing project" was introduced to cater low-income urban families with housing problems, with selling prices just covering the basic construction cost. And three urban housing categories - commodity, partial "welfare" (sustained by danwei), and social security housing - coexist to constitute a transitional housing system. [76], p. 515)

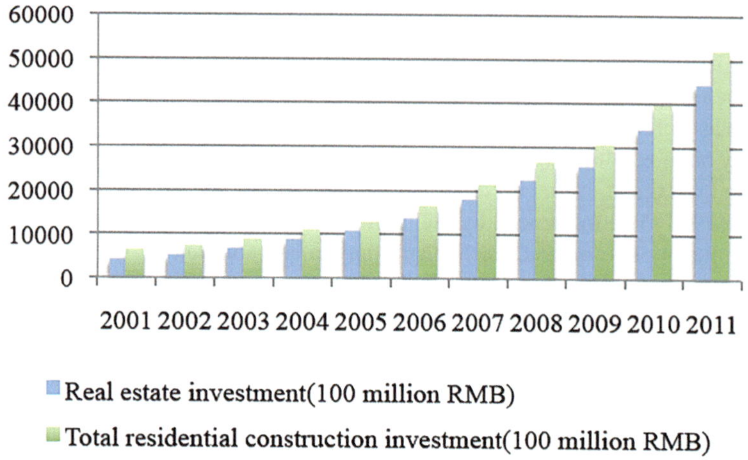

Fig. 3.1 Real estate investment and total residential construction investment. *Source* China Statistic Year Book 2012

During the fourth stage of post-2001 period, in light of housing commodification and the emphasis on real estate industry as a growth anchor in response to the Asian financial, the past decade has witnessed the proliferation of commodity housing estates [36, 59, 62]. The real estate investment increased yearly from 2001 to 2011, and its proportion in total residential construction investment has risen from 67.3% in 2001 to 85.6% in 2011 (Figs. 3.1 and 3.2). China's housing provision finally turned into an open and private market system and the housing tenure transformation has accomplished [6]. Instead of relying on the welfare housing provided by the state or state-owner enterprises (*work unit*), most families had to turn to the housing market to satisfy their housing needs [6]. Thus, commodity housing is important for urban residents, which is developed by market-based real estate developers, planned by designers or architects and managed by property management companies [62]. Furthermore, the high material standard of living and new freedoms to create a private life are brought by marketization [11]. So developers advertise high standard of living in a comfortable space [74], especially focusing on the large housing estates or *xiaoqu* that primarily cater to the nouveau *riche* and the new middle-class [36].

3.1.1.2 Trend of Nurturing Suburban Living

Accompanying with commodity housing development, urban expansion and decentralization have been underway simultaneously. To house the increasing urban population, cities have to expand outwards rapidly, undergoing spatial and economic growth. Suburbs have thus become an important arena of urban development [64], and residential suburbanization is largely promoted by the commodity housing development.

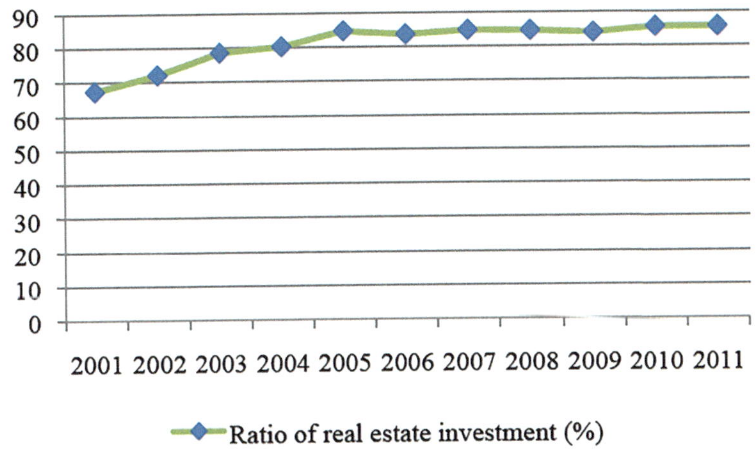

Fig. 3.2 Ratio of real estate investment in total residential construction investment. *Source* China Statistic Year Book 2012

Economic, land and housing reforms have contributed to residential suburbanization. Stimulated by the economic reform, opening-up policy, land and housing reforms, urban growth, and rising incomes, suburbanization grows rapidly in China [15]. Notably, land reforms contribute to the sustained urban spatial expansion to urban fringes, where the cost of land remains low, leading to the speedy suburbanization [16, 75]. Local governments also contributed to rapid suburban development (Fleischer [16], as cited in Yan et al. 2002). To stimulate local economic growth and autonomy, and to enhance competitiveness consciousness, large land plots were converted into construction land in suburban areas, and different forms of suburbanization have appeared in residential, industrial and retail development [65, 75]. With the development of the land market, there is great impetus for local governments and real estate developers to develop commodity housing in suburbs. Specifically, the large-scale commodity housing estates are in the interests of all parties, including local government, real estate developers, and consumers [45], thus accelerating the pace of suburbanization, creating a new lifestyle choice.

With the increase in motor traffic, households began to look for a better environment and opt for a lifestyle based on automobile [14]. From the mid-2000s, the concept of suburban living is wider accepted, further accelerated the development of commodity housing in suburbs [65]. In addition, the abolition of administrative allocation of housing and the further development of housing market led to the freedom and flexibility in selecting housing location, and housing prices and the living environment become crucial factors making in housing choices [14, 74]. Thus, pushed by the housing market and pulled by the desire for a better living environment, a preference for suburban commodity housing gradually emerged. Accordingly, SCHEs become a new urban landscape and lifestyle choice in Chinese cities. These estates are featured with and advertised for their lower housing price (relative to housing in central city location), better living conditions and greener environment. Thus, more affluent families are found in the SCHEs, compared with the manufacturing workers and rural migrations and the low-middle income urban residents for the affordable housing in suburbs [50].

To summarize, consequential to economic, housing and land reforms, commodity housing development has expanded to suburban areas. The development of property market has provided a basis for developing commodity housing estates in suburbs, and modern transportation and information techniques accelerated the process [16]. This has led to a higher share of suburban housing. The increasing attractiveness of SCHEs implies that more and more middle-income population move to suburban living, the health concern of this living style thus needs to be investigated has been generated for the middle-income population.

3.1.1.3 Suburban Commodity Housing Estates (SCHEs): The Focus of This Study

Based on previous discussion, this study focuses on the suburban neighborhoods in China, paying particular attention to suburban commodity housing estates (SCHEs). These distinctive forms of human settlements and the resultant styles of suburban living inevitably affect residents' health, which has been rarely discussed in the existing literature. Detailed reasons are discussed as follows.

First, the SCHE is a typical form of suburban neighborhood, whose development has promoted residential suburbanization greatly, especially in Guangzhou. The market share of SCHEs in housing market is high and SCHEs have great impact on suburban development and urban growth. In addition, compared with the images of low rise and dispersive distribution of US model of suburbs, it is an emerging phenomenon that the middle-income communities in suburbs are more densely populated [48]. SCHEs in China contribute to the high-rises and compact landscape of suburban neighborhoods, where villas, townhouses, multi-story and high-rise buildings make up the SCHEs.

Second, SCHEs are typical of suburban neighborhoods. Other than high-density land use configuration, the leapfrog development pattern and gated enclaves are two other identifiable features. For one thing, influenced by the urban expansion to the periphery, the land enclosure and the improved intra-urban transportation, the discontinuous form has dominated the suburban space [65]. Accordingly, a form of leapfrogging peripheral growth resulted, and SCHEs have contributed to the discontinuous suburban form, comprised of residential enclaves [3, 33, 65]. Most SCHEs are gated communities and privately managed, separated from the surrounding urban matrix. Compared with work-unit compounds, the gated suburbia is differentiated by the new way of life, which is symbolized by the features of packaging and branding [63]. These enclaves are not only distinguished by their spatial form, but also by their social characteristics and their mode of neighborhood governance [3].

Third, territorially, SCHEs are located at the peripheral area and transitional areas beyond the urban core area but within the metropolitan area, reflecting the attributes of suburbs. Although the suburbs are situated on both urban and rural space and are administered by both urban and rural institutions, the development of SCHEs is urban-oriented. Located in the urban periphery, the SCHEs are physically separated from the central city but depend on it economically and socially, especially in terms of employment [69]. In this context, commuting between the urban central are and SCHEs is necessary for the daily life of residents, and the public transportation system has been developing to accommodate the increasing traffic demands. However, it has been argued that development of SCHEs has been driven by the boom of using of private vehicles to some extent.

Forth, SCHEs are socioeconomically homogenous, skewing forwards middle-income. Due to urban gentrification, the majority of middle-income residents are unable to afford living in city centers, so most of them settle in neighborhoods of high-rise apartments on urban periphery [48]. Moreover, SCHEs are enclaves in terms of their social status, and the attribute of packaged residential enclave makes the SCHEs attractive to certain population, which may represent something of a backlands utopia [69]. It is observed that the SCHEs are mostly designed according to the tastes of urban middle-income (mid-level managements, executives and professionals), who have sufficient incomes to pay market prices for their home [3, 33, 69].

Therefore, the SCHEs are selected as the focus of this study, featured by peripheral middle-income gated enclaves with relatively high density and a leapfrog development pattern. The development of SCHEs in Guangzhou is then reviewed.

3.1.2 SCHEs Development in Guangzhou

3.1.2.1 Guangzhou's Leading Role in Suburban Housing Development

Guangzhou is selected as the empirical case due to its leading role in the development of SCHEs. With a recorded history of 2800 years, Guangzhou is situated at the northern end of the Pearl River Delta, facing the South China Sea [9]. It is a subtropical city, characterized by the abundant rainfall and perennial verdure all year around. As the capital of Guangzhou Province, it is a political, economic, scientific, technological, educational and cultural center, comprising 11 districts with a population exceeding 12.84 million in 2012 within a total area of 7434.4 km^2 [23].

After 1978, Guangzhou pioneered the China's economic restructuring and urban spatial restructuring, land reform, urbanization and suburbanization, as well as charting a new path for housing development [8, 65]. "The rich and entrenched commercial history and geographic proximity to Hong Kong naturally accorded Guangzhou special positions in the reform period" [56]. So market elements and foreign investments infiltrated Guangzhou in the early reform period, especially the inflow of capital from Hong Kong in the mid-1980s (Wang and Li [56]), contributing to the spread of commodity housing. At the same time, promoted by the autonomy in the wake of political decentralization in 1979, housing reform progressed rapidly in Guangzhou, accelerating housing commodification and marketization [26]. Thus, housing market in Guangzhou is not only the most developed one in China but also typical for other cities to refer [27, 65]. Tactics such as using land-use rights to stimulate developers to construct road and mass transit and large residential neighborhood development in suburbs have been learnt by other cities in China [65].

Promoted by housing commodification and urbanization, Tianhe, Haizhu, and Panyu districts in Guangzhou were under rapid housing development, leading to the expansion of residential areas to the east and south of Guangzhou. Panyu District, a new district of Guangzhou, has been undergoing residential suburbanization for three decades, making it a good example for investigating the effects of suburban living on residents' health. Three periods of suburban housing can be identified in Panyu from the institutional perspective.

3.1.2.2 Stage 1: Market-Government Cooperation Period

The late 1980s to late 1990s constituted the first stage of residential suburbanization in Guangzhou, during which Panyu has witnessed speedy housing development. From 1990s, promoted by the policy and process of housing privatization, housing estates developed expeditiously in Guangzhou, expanding southward to Panyu district and forming a sprawling, multi-level residential suburban agglomeration [23]. Affected by the first administrative adjustment of Panyu in 1992, which changed Panyu from a county into a county-level city affiliated with Guangzhou, decentralization brought the emergence of large commodity housing, such as those in Luoxi area (*Luo xi ban kuai in Chinese*): Luoxi New Town (1989, 150 ha) and Lijiang Garden (1990, 81 ha). With the completion of Luoxi Bridge in 1988 and Huanan Express Way in 1999, the traffic bottleneck between Guangzhou city and Panyu District has been broken, triggering housing development in Panyu.

In the context of uncompleted economic and housing reform, the initial stage of commodity housing development has been characterized in terms of cooperation between public sectors and developers, especially referring to foreign-related investment. This mode has been inaugurated by the first commodity housing estate in China, Donghu New Estate, built by the cooperation between the public sector[3] of Guangzhou and a Hong Kong developer.[4] This mode of cooperation between public sector and private sector has been adopted in the SCHEs. Specifically, the housing development in Panyu in the first stage was promoted by "Panyu Real Estate Joint Development Company",[5] which is a government-linked company with a specialization in urban development. For example, three housing estates: Luoxi new town project (LNT), Clifford Estate, and Lijiang Garden, have been developed by joint ventures 'Luoxi Real Estate co., Ltd.', 'Clifford Group', and 'Guangdong Holdings Limited', respectively, all of which are based on cooperation between the 'Panyu

[3] It refers to the 'Dongshan district residential construction headquarter of introducing foreign capital' (东山区引进外资住宅建设指挥部).

[4] It is the 'Chrysoberyl River Development ltd'.

[5] In Chinese, it is referred as "shi fang di chan lian he kai fa zong gong si" (市房地产联合开发总公司)

real estate joint development company' and Hong Kong developers. For the sake of housing development, the government-linked sectors usually provide assistance with land development and development while the private sector provides funding.

In terms of the characteristics of the SCHEs, the targeted population has significant influence. Two groups of population have been designated as the potential buyers: the working class in Guangzhou and Hong Kong people from the export market. In order to attract consumers, developers have emphasized the concept of suburban living [60]. As a result, a range of neighborhood types has been developed to satisfy housing demands from villas to high-rises, with diverse layouts including both fine-grid open street distribution and large scale gated communities. Additionally, housing management includes both extensive management and gated property management derived from HK. Moreover, the comprehensive development mode has been adopted for the large-scale housing estates. Developers would invest and construct public facilities before the completion of the whole housing projects, such as schools, retails, and community service facilities [35]. In addition, the affordable housing price and relatively good environmental quality are the attractive elements as well [37].

To summarize, suburban housing development in this period is pioneered by the cooperation of state and market, on setting the residential suburbanization in Guangzhou.

3.1.2.3 Stage 2: Developer-Dominated Period

The second stage of suburban housing development in Guangzhou could be designated as from late 1990s to the late half of the first decade of the twenty-first century, featured by the domination of the developers in the development of large-scale packaged housing estates. With ongoing housing commodification, the booming of real estate industry, and the stimulus from the second administrative adjustment in 2000 annexing Panyu city into to a district of Guangzhou, the process of "crazy" land enclosure has been triggered, resulting in the "Dapan period[6]"of the property market. With the releasing of housing demand, large pieces of cheap land in suburban areas, and the loose planning control, developers and land speculators prefer to choose large sites for residential developments in suburbs [69]. By 2002, more than 30 housing estates in Guangzhou have more than 66.7 hm^2 (equal to 1000 mu) of land [38]. In Panyu, eight large-scale housing estates are quite famous during this period, developed from the late 1990s or the early 2000s. Due to the large scale, these housing estates are comprised of different clusters, which are developed in chronological order over the years [57].

In the second stage, housing reform has intensified housing commodification, suggesting the increasing impact of the property market on SCHEs development.

[6] Dapan is a Chinese word, referring to large-scale housing estates, which are usually branded by famous real estate developers. It could be seen as a term developed by developers.

Stemming from the dominance of developers and the absence of government intervention, new residential communities in suburbs are often built into residential enclaves, where they offer their own public services [3, 33, 65]. Many studies on the commodity housing in China focus on this period, revealing several characteristics of the suburban housing estates. First, most suburban housing estates are privately managed gated communities, separated from the surrounding urban matrix, especially the workplaces. Although the community atmosphere of gated communities is not new compared with the work-unit compound, the new way of life in the gated suburbia is new, which is symbolized by the packaging and branding attributes [63].

Second, under China's economic reform, service delivery has been reorganized, shifting the consumption pattern towards club realms (Wu [63], as cited in [62]). Due to the lack of previous residential development in suburbs, insufficiency of public facilities, and high cost of improvement, developers resorted to packaged development to provide services such as sanitation, education, medical care, security and transportation, in addition to a congenial and socially acceptable neighborhood [33, 69]. With the lack of local public services, developers have to provide "shuttle bus, schools, hospitals, estate security and fire protection, sewage disposal and even waste water treatment" (Li [34], p. 95). Thus, with developers becoming responsible for providing service facilities, the strategy of "packaged" service facilities has been used by developers in commodity housing development, emphasizing both high environmental quality and comprehensive, high-quality services [62]. This leads to the reorganization of service delivery from the consumption pattern to club realms [63]. In order to capitalize upon market opportunities, developers advertise their products in many ways, such as branding and cooperation with prestigious primary and secondary schools. Thus, the packaged enclave mode is the developers' way of providing service facilities to residents that should have been the responsibility of the government. In this sense, the suburban housing estates are featured by diversity, packing, branding and individualization [25, 63].

In short, developers dominated residential suburbanization by providing diverse housing products and public facilities, while the government transfers the responsibilities of providing public service to the developers. This leads to the packing, branding and individualization of suburban housing estates.

3.1.2.4 Stage 3: Planning Regulation Enhanced Period

The period approximately from the late half of the first decade of the twenty-first century until now is referred as the third stage, with the stricter land leasing for real estate development by the Guangzhou municipal government. Few studies related to commodity housing have been conducted in this period, and the characteristics of the suburban housing estates with the tightened policy control of the property market since 2010 remain unknown.

Constrained by limited literature in this period, the author conducted some empirical research based on the data from Municipal Land Resources and Housing Administrative Bureau in terms of land use approval for residential use are collected from

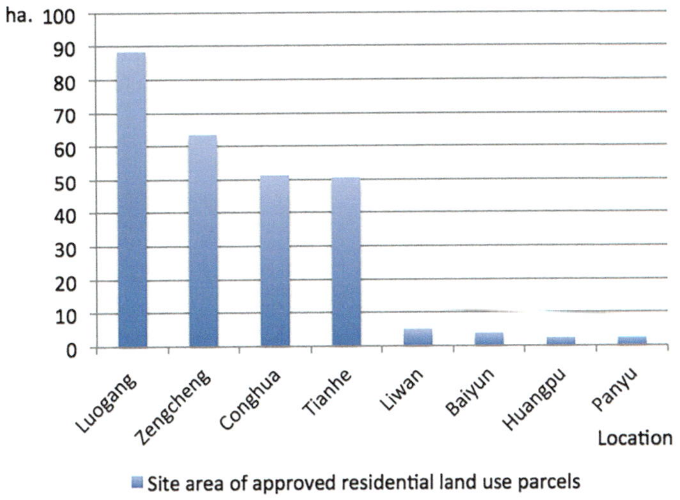

Fig. 3.3 Distribution of approved residential land use parcels in Guangzhou (07/2011–01/2015)[8] *Source* Official website of Guangzhou Municipal Land Resources and Housing Administrative Bureau

July 2011 to January 2015. Nineteen land parcels have been approved for residential use,[7] 79.2% plots of which are in suburban area, including Luogang, Zengcheng, Conghua, Baiyun, Huangpu, and Panyu (Fig. 3.3). The majority of these plots are in outer suburbs: Luogang, Zengcheng, and Conghua, accounting for 76% of all approved residential land parcels. This could mean that most new housing estates concentrate in the outer suburbs of Guangzhou, while few are in Panyu. In addition, nine land parcels approved for residential land range from 0.77 ha to 50.54 ha, five of which (26.31%) exceed 20 ha and seven (36.84%) exceed 10 ha (Fig. 3.4). It is possible that the site scale of the approved residential land after 2011 is much smaller than the housing estates in the second stage. For land use, half of the parcels (9 plots) are designated not only for housing but also for green space, roads, infrastructures, commercial space, schools, hospital or warehouse. For the five plots exceeding 20 ha, only one plot has been approved for net residential land use, while more than two-thirds of net residential use plots (70%) cover less than 10 ha (Fig. 3.5). Thus, this could mean that the scale and function of residential land use, especially net residential plots, are under tight government control.

Thus, planning has become more strictly controlled than in the previous two stages since the Regulation Plan of Guangzhou (full coverage) was introduced in 2011. Constrained by the size of land parcels, developers prefer to produce fewer types of housing than they did in the first and second stages, focusing on high-rises especially. In addition, unlike the large-scale housing estates, small or medium ones

[7] Except the land for affordable housing.

[8] Source: Official website of Guangzhou Municipal Land Resources and Housing Administrative Bureau

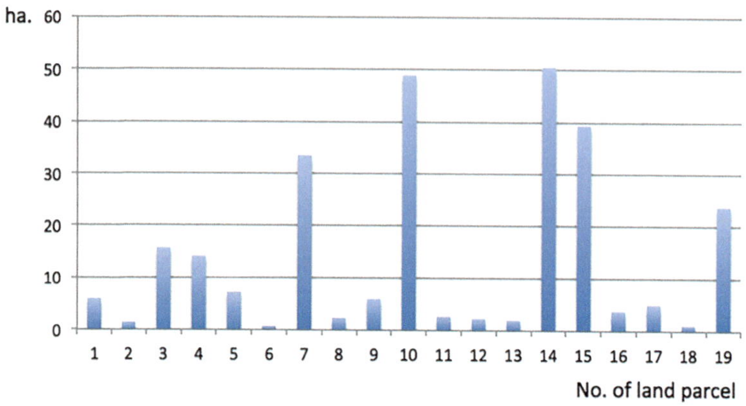

Fig. 3.4 Site area of approved residential land use parcels in Guangzhou (07/2011–01/2015) (ha.)[18] *Source* Official website of Guangzhou Municipal Land Resources and Housing Administrative Bureau

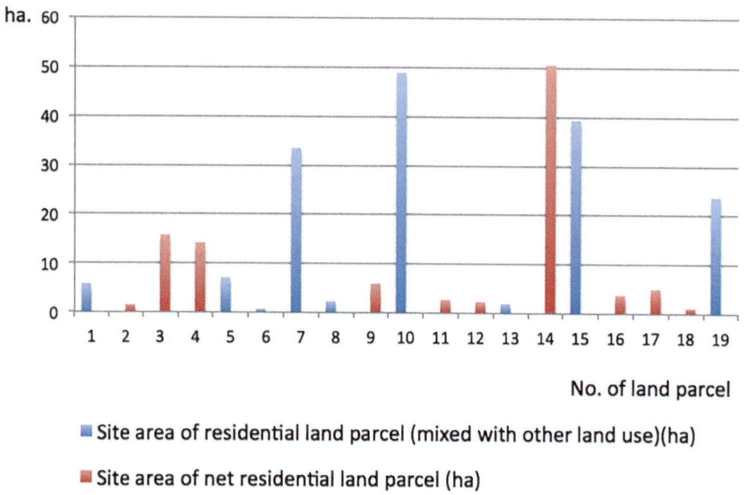

Fig. 3.5 Land use pattern of the approved residential land use parcels in Guangzhou (07/2011–01/2015)[18] *Source* Official website of Guangzhou Municipal Land Resources and Housing Administrative Bureau

tend to aggregate to share public facilities. Individual developers no longer have to provide all the day-to-day amenities. However, developers still need to construct public facilities like schools and infrastructure facilities according to the terms of their lease.

3.1.3 Issues Against Healthy Neighborhood in SCHEs Development

This section identifies empirical issues during the development process that may undermine a healthy neighborhood. Although the quality of the living environment in SCHEs is good due to their marketed-based development and the peripheral location, the impacts of the SCHEs development on health performance of residents have received little attention. Issues in SCHEs development that can impede the creation of a healthy neighborhood may derive from discrepancies between market-driven and people-oriented principles of development, and discrepancies between market demands and health needs.

3.1.3.1 Market-Driven Rather Than People-Oriented Development

The planning and design of the SCHEs are based on the market conditions rather than people-oriented consideration. With the property market booming, market forces have become more influential in residential planning and design since 1990s. With the developers' interest in satisfying a variety of market demands, residential planning and design has transited to market-oriented. This change could be a double-edged sword, which could advance the living standard rapidly but could be at the cost of people-environment congruity. In this context, the planning of SCHEs is influenced by market demands because of the leading role of market-based real estate developers [62]. Thus, the making of residential space in China has gone beyond providing accommodation for the workforce and is becoming a drive of the economy [63].

To capture the market, SCHEs are often advertised as a lifestyle by developers with advertising gimmicks. In the highly competitive real estate market (especially for upper-income housing), developers are facing two interrelated process of marketization and aestheticization [63]. Marketization raises the demand for aesthetically appealing goods, while the creation of an aesthetic living environment needs to overcome the bottleneck of commodity housing development [63]. Developers believe that the particular lifestyle and the physical form of residential environment could eventually transform Chinese society, and a slogan they put forward is living transforming China (*Juzhu gaibian zhongguo*) [63]. Housing is representative of lifestyle for both consumers and developers [58]. In practice, developers usually borrow exotic concepts to advertise their housing products with the recognition of the importance of packaging and branding [63]. To fulfill the desires for good life, developers use the concept of "exotic" and "stylish" new living space to produce the new gated and packaged suburbia [63]. To operate, most SCHEs are gated communities with spatial enclosures such as secured gates, walls, fences and security personnel [62, 63]. The gate sends a message of desirability about residential areas, marketing the housing projects as the entrance to the good life [63].

Additionally, the dominance of market force interferes with the provision of public service negatively. The public services are replaced by privately paid-for services in the packaged suburbs [63]. As one example, shuttle buses are subsidized and financed heavily by developers for home sales, so developers usually reduce the subsidy when the homes are sold out [34]. This way of operation could produce a vicious circle and it will be terminated at last [34]. Additionally, the market-based strategies for service facilities provision highlight discrepancies between consumers' demand and health needs, threatening the prospects of healthy living in the SCHEs.

Consistent with the trend of aestheticization, developers are obsessed with providing large, green outdoor open spaces and great landscapes in SCHEs, and a large part of investment has been diverted into landscape construction to attract more buyers [23]. For developers, a significant job is to make the landscape of neighborhood looks nice and aesthetic [63], and they would only actively provide green space when they believe that it could be beneficial for their competitiveness in housing market [71]. As a result, a majority of residential landscapes endure low biodiversity, low usage rate, but high energy and resource expenditure in maintenance [23]. Thus, focusing on how it looks rather than how it works and how it is used, the landscapes of the SCHEs are mostly designed to please consumers visually [23].

In summary, given the market-oriented development of SCHEs, it has been criticized that although developers are aware of the buyers' fondness for high-quality environments increasingly, they are still lack of knowledge of local housing preference (as cited in [27, 59]). There is still a gap between the housing supplied and what consumers need [27], especially in terms of health interests. Thus, it is concluded that planning and design of SCHEs have been driven by market forces at the expense of people-oriented benefits and people-environment congruity.

3.1.3.2 Discrepancies Between Market Demands and Health Needs

With market-driven development, gaps inevitably appear between the residential environment provided by the developers and the health-promoting environment, formed by the discrepancies between market demands and health needs.

In terms of market demand, different housing conditions induce different housing demand and preference, and the persistent improvements engender feedback to promote housing demands to high-order expectations for environmental quality [27]. After more than two decades of housing reform and marketization, improving the living environment is the major concern of consumers [27]. Based on Jim and Chen's [27] survey with households in the new residences sold in urban areas of Guangzhou in 2004, the home-buyers' preferences for outdoor environmental attributes have been examined. Eight key factors and ten ancillary factors determining housing consumption are identified. Among the key factors, price is the predominant one, followed by *the need for security*, which could be measured by gated communities, with perimeter walls and fences, secured gates, security personnel and professional property management [27, 62]. *Quality property management*, associated with the need for security, is third. Other key factors include *location, accessibility and mobility*, particularly proximity to shopping area, workplace and public transport

[27]. The last two key factors are *environmental elements within and around residential precincts*, comprised of five elements (green space, open space, environmental quality, water body, and children's playground), among which green space and environmental quality are the most important demands of home-buyers [27]. The ancillary factors are less important than the key factors in consumers' preference, although most respondents rated them as very important or important. In descending order, ancillary factors are identified as: "quality of building and decoration, proximity to school, provision of public facilities, reputation of the developer, design of rooms in the apartment, and design of building and development, proximity to metro station, popularity of the development and location in old towns and new towns" ([27], p. 422). Combining the key and ancillary factors, the determinants of consumers' preference in buying commodity housing estates are unveiled, which also could be seen as the demands for a satisfactory living environment.

As demonstrated in Chap. 2, parameters for healthy neighborhoods could be cited in determining the health needs of residents of housing estates. Determinants of health at the neighborhood scale are divided into three aspects: environmental factors, personal factors, and interrelations between people and environment. Generally, a healthy neighborhood should promote physical activity and reduce the risk of obesity and overweight, promote mental health and reduce environmental stress, promote social support and neighborhood attachment. Most epidemiologists acknowledge that the built environmental factors that affect eating and physical activity behavior and social interactions are independent predictors of health [49]. For example, density or intensity could increase the access to service destinations, the length and frequency of trips, the demand for walking and cycling, car dependence, community participation and social capital [49]. Similarly, mobility (walkability, accessibility and convenience), connectivity, access to public transit facilities and car-dependence degree, diversity and job-house balance, and people-oriented design could all have effects on the physical activity behavior and social interactions. Therefore, health needs could be interpreted as: relative high-density, high walkability, good accessibility to destinations (retail, green space, healthy food provision, health care facilities, educational institutions, etc.), easy connection between trip origins, active travel, proximity to public transportation, mixed land use, adequate and quality greenness, opportunities for physical activity, social support and interpersonal contacts, safe and aesthetic neighborhood [19, 30, 49, 53].

Comparing the consumers' demands with health needs, several discrepancies could be found. Table 3.1 summarizes the market demands based on the study of Jim and Chen [27] and aggregates the health needs based on the literature review in Chap. 2, particularly drew from the health-promoting environmental characteristics demonstrated in Sect. 2.3.1. In Table 3.1, the terms in blue (price, quality property management, reputation of the developer, popularity of the development) have little relevance or indirect relevance with health determinants of the residential environment, but are important in housing choice. The terms in red (high-density, high walkability, easy context between trip origins, active travel, mixed land use, opportunities for physical activity, social support and interpersonal contacts) have

Table 3.1 Comparison between market demands and health needs

Market demands [27]	Health needs (Summarized by author)
• Price • Security • Quality property management • Location • Accessibility • Mobility • Green space and environmental quality • Quality of building and decoration, • Proximity to school and metro station • Provision of public facilities, • Reputation of the developer, • Design of rooms in the apartment, building and development • Popularity of the development	• High-density • High walkability • Good accessibility to destinations (retail, green space, healthy food provision, health care facilities, educational institutions, etc.) • Easy connect between trip origins • Active travel • Proximity to public transportation • Mixed land use • Enough and quality greenness • Opportunities for physical activity • Social support and interpersonal contact • Safe and aesthetic neighborhood

received little attention in housing preference, although they bring benefits to residents' health. Thus, these discrepancies show that homebuyers usually pay more attentions to property value to their health when purchasing commodity housing. In addition, although there is some overlap between market demands and health needs, they could have different foci. For example, the safety aspect emphasizes the gate in market demands while focusing on risks of injury in health needs.

3.2 Health Issues of Suburban Living

In terms of urban living and health, it is acknowledged that urban living is associated with heightened risks of chronic disorders, a demanding and stressful social environment and great social disparities [31]. In the dynamic development of residential suburbanization, it is supposed that residential living in SCHEs may have great effects on health. To support this proposition, this section ascertains the health issues stemming from suburban living through a review of related studies.

3.2.1 Widely Discussed Issues in Suburban Living

There are several widely discussed empirical issues of suburban living in Guangzhou, focusing on living in the large-scale housing estates developed during the first two periods. First, it is acknowledged that living in the SCHEs depends on the urban core area heavily, especially for job opportunities, and the SCHEs are excluded and separated from the host suburbs at the same time [34]. Since tertiary employment and higher-status services congregate in urban core areas, most residents of suburbs commute to their jobs in urban core areas daily [34]. Accordingly, problems associated with commuting emerged, such as time-consuming long commutes and rush-hour congested traffic [72].

Another acrimonious issue is the lack of urban services and amenities in the host area for suburban residents. Due to the spread of the suburban housing, local provision of public service has lagged behind the increasing demand [34]. Although developers provide service facilities in large housing estates, the supply cannot meet the demand of residents, especially the daily commuting, medical care and education [39]. Notably, the lack of public transportation services has influenced the residents' daily travel greatly. According to Lin et al. [39]'s survey in 2007, 71.6% residents in Panyu District used public transportation for daily travel. Besides the long commuting journeys and high fares, suburban residents have to spend higher living cost especially for getting proper medical care and quality education [34]. Moreover, it is criticized that the packaged and gated development of large housing estates has brought the deterioration of surrounding areas and duplicated construction of service facilities [72]. Both long commuting and insufficient urban services would have health impacts, which need to be further investigated.

3.2.2 *Threats of Obesity*

3.2.2.1 Obesity in China

"Increasing obesity and physical inactivity are problems across the world, and especially where rapid industrialization and urbanization have occurred" ([1], p. 123). Increasing statistical evidence suggests that obesity has become an urban issue in China since the early 1990s [17]. By 2004, nearly a quarter of all Chinese adults were overweight, and the proportions of overweight and obesity and physical inactivity are rising [7]. The rates are expected at least to double over the next two decades [46]. The WHO report of 1989 suggests that China was thought to be lean, but it is catching up the West with regards to the incidence of obesity [66]. The overall prevalence of overweight and obesity of Chinese population increased by 38.6% and 80.6%, respectively, from 1992 to 2002, indicting a remarkably rapid transition [41]. Data from the 2002 China National Nutrition and Health survey show that the overweight prevalence was 22.8% (BMI $\geq$ 25 kg m^{-2}) and for obesity 7.1% (BMI $\geq$ 30 kg m^{-2}) [5]. As a result, the spread of obesity contributes to the increase of chronic disease and cardiovascular disease, such as the risks of hypertension, diabetes, stroke and coronary heart disease [5]. Additionally, obesity imposes as a substantial economic burden on China, accounting for 4.5% of national health expenditure in 2010 [70].

Explanations for the increase of obesity and overweight include diet changes, reduced energy expenditure caused by reduced level of physical activity, increased sedentary lifestyles, and inactive occupations [1]. Physical inactivity has been identified as one of the major public health problems of the twenty-first century [2,

55]. In China, physical inactivity and sedentary lifestyle are prevalent in adolescents [7, 55]. However, the significance of physical activity has been underrated and underappreciated in public health and urban planning [2].

3.2.2.2 Suburban Living and Sedentary Lifestyle

In reality, the prevalence of obesity across the living locations is proved to be different, such as that the prevalence of obesity was higher for urban residents than residents with township and rural status [77]. This phenomenon could be supported by the discourse on influential factors that promote the decline of energy expenditure in the literature, such as that urbanization, urban expansion and industrial development contribute to the increased acceptance of unhealthy lifestyles such as reduction of active transportation and a decline of occupational physical activity [1]. Thus, the urban environment could encourage a sedentary lifestyle.

Lifestyle choices have a direct impact on residents' physical and mental well-being [32], among which dietary intake and physical activity are the major aspects [67]. As a general way of living, Chinese urban residents' lifestyles have changed considerably over the last 30 years due to the urban spatial restructuring and industrial transformations [17]. Accordingly, there is has been a speedy shift in diet and physical activity, subsequently resulting in obesity and related noncommunicable diseases, and nutrition-related causes of death [17, 46]. Thus, the expeditious urbanization in China leads to the popularity of urban lifestyle, which is more sedentary, more dependence on automotive transport, more inactivate jobs, more time sitting at rest and play, more discretionary income available to spend on treats [17]. Obstacles in Chinese cities reducing the level of physical activity are identified by some research, such as the barriers and inadequate sidewalks for walking, single land use pattern, and policies supporting car production and road building, and even the cultural recognition of affluence and success that are associated with car ownership [1].

It is argued that the SCHEs are conducive to a sedentary lifestyle. The early genesis of an American-style, 'Starbucks-and-Saks' style suburban culture is emerging and developing in China recently [17], so the empirical research in US context could be mentioned. Studies conducted in the US might confirm the associations between urban form and forms of physical activity, which affect physical functioning, quality of life and overall health [28]. An overall decline in total physical activity is found with a strong and consistent trend towards Americans living in suburbs [4]. For example, it is found that people living in the sprawling environment intend to walk less, weigh more, suffering from higher prevalence of hypertension than those in compact urban environment [12]. In addition, automobile-oriented lifestyle contributes to a less activity-oriented society [4]. Thus, the urban sprawl threats residents' health through encouraging the sedentary lifestyle, which increases the risk of cardiovascular disease, stroke, diabetes, and obesity [19]. Although Chinese suburbanization is different from Western suburbanization, they both are associated with a decline in physical activity.

Choosing a house means choosing a lifestyle [15], particularly when people purchase SCHEs. Living in the SCHEs means long commutes, and residents depend more on motorized traffic rather than walking or bicycling [17]. Simultaneously, the long commutes leave little time for exercises after work, and consequently, exercise rates have plummeted because of the changing lifestyles, lack of time and distance from facilities [17]. For the changing of lifestyle, moving to suburbs has broken up previous social networks of residents, and residents tend to spend nights in front of the TV or the PC instead of social gathering and evening stroll [17]. For the diet, increasing people rely on convenience foods from convenience stores on their commuting routes [17]. In addition, the options of public facilities are limited, especially for the educational and health care facilities, implying that inefficiency of leapfrogging development and packed enclave development increase the commutes to urban core areas [17]. This situation discourages physical activity, suggesting that the popularity of suburban living could have contributed to the rise of proportion of physical inactivity in day-to-day life of Chinese citizens.

As a result, as a new middle-class, residents in SCHEs are potentially facing the threats of obesity, which is associated with the risks and incidence of chronic diseases (capsular disease, cancer) particularly [17]. Obesity has become a key middle-class issue in China, and the middle-class is getting fat fast because they are abandoning their traditional diet and lifestyle [17]. Specifically, the wealthy and newly emergent middle-class consumers have specific ideas about their living environment and lifestyle, and they choose to live in the newly expanding cities for their green space, cleaner air quality, spaciousness, privacy and exclusivity [15, 17]. This home choice may increase the possibility of living in a sedentary way. Thus, it is proposed that the middle-income residents in SCHEs may not be physical active enough due to the suburban living lifestyle, which may decline energy expenditure and thus increase the obese-related diseases.

3.2.3 *Threats of Mental Disorders*

3.2.3.1 Mental Disorders in Urban Environment

Mental disorders are found in all regions, countries and societies, among women and men at all stages of the life course, rich and poor, and urban and rural residents [61]. Around one in three people has at least one psychiatric symptom, though it is mild such as tension, anxiety or depressed [21]. It is estimated that depression and depression-related illness will be the greatest source of ill-health by 2020 [47]. In China, mental disorders have been largely ignored, especially compared with physical health, leading to the increasing burden of mental disorders [61]. The prevalence of mental disorders in China has been controversial for decades, because of its lower estimated prevalence than those found in psychiatric epidemiological surveys elsewhere in the world [51, 54]. For this controversy, several potential reasons are proposed, including methodological differences, conceptual and measurement problems, stigma induced conscious non-disclosure, and even the broad culture characteristics [51].

It is acknowledged that mental health is negatively affected by cities: higher prevalence of mood and anxiety disorders are associated with city dwellers, and higher incidence of schizophrenia is also associated with people born and raised in cities [31]. Concerning urban environment, variations in prevalence of mental disorders between different urban neighborhoods have been studied since the 1930s [10]. Notably, focusing on urban–rural differences in prevalence ratios of mental disorders, meta-analyses show that city dwellers have a 21% increased risk for anxiety disorders and a 39% risk of mood disorders controlling for important confounders [44]. In the metropolitan Chicago, the prevalence of mental disorders decreases from the urban center to the outskirts, indicating a close correlation between mental disorders and the ecological structure of the city [13]. In addition, the relationship between mental disorders and urban living has been addressed in many studies. Most research attributes mental disorders to urban social environment, and they affirm that the causation mechanism is more important than the selection mechanism, which supports the argument that urban environment causes psychosis [29]. Emerging studies pay special attention to the neural process, which could mediate the unidentified associations between mental disorders and the urban environment [31]. Using functional magnetic resonance imaging in three independent experiments, [31] find noticeable neural mechanism that could link the urban environment to social stress processing, and they argue that urban upbringing and city living have divisible effects on social evaluative stress processing in humans.

However, in China, the relevant studies are much fewer and whether and what effects urban environment has on mental health remain unclear [20]. Empirically, only some studies focus on the mental conditions of the rural-to-urban migrants who are excluded from the urban public service and blocked from access to public health due to the hukou system [24]. Thus, there is still a gap in the empirical research on the topic of mental disorders in urban China.

3.2.3.2 Suburban Neighborhood and Mental Health

Based on the review of the inherent connections between mental disorders and the urban environment, it is logical that suburban environment may have both health risks and benefits. But most studies focus on the negative effects of suburban living on mental health.

The original studies on mental health in new suburban area were conducted in Great Britain as early as 1930s [10]. In the first studies, neuroses were found to be more prevalent in the new suburban areas than in inner city due to the loss of family surrounding, long distance from job, and social isolation (Dalgard and Tambs [10]). This finding is supported by a later study by Martinet al. [42], who find a high incidence of psycho-neurotic disorders in population rehoused in suburban areas and propose the loneliness and social isolation as important explanatory factors.

However, some studies conducted in 1960s found no significant difference in incidence of neuroses between the suburban new town and the older city, the prevalence of psychoses in suburban area is much lower (Dalgard and Tambs [10]). Accordingly, a proposition has been raised: "psychoses are more influenced by the environment than had been previously thought, whereas neuroses were more dependent on constitutional factors". In contrast, based on a comparative study, Hare and Shaw [22] argue that no substantial differences could be found in mental disorders between housing estate in suburbs and those in the heart of the city. Additionally, the spaciousness and greenness of suburbs are regarded as the healthy corrective to the psychologically distressful urban experience in some studies (Oliver [43]).

In the US, with the rapid and steadily suburban growth, half of the population lived in the suburbs at the beginning of twenty-first century, triggering a discussion of urban sprawl and public health in the academic literature. The relations between suburban environment and mental health remain a matter of debate. Although the public discourse considers the suburbs as psychologically unhealthy places in the US, there is no sufficient evidence of an authentic link between the suburban environment and psychological distress [43]. Empirically, focusing on city-suburb dichotomies, the relationships between metropolitan places (population size, density, racial composition, affluence, age, land use) and mental health (depression, life satisfaction, self-efficacy, esteem) are examined. It is proposed that high density and less affluence lead to high prevalence of depression and dissatisfaction; however, suburbs could not be considered more psychologically damaging than cities [43]. In addition, no substantial associations have been found between mental health outcomes and sprawl environment. Sturm and Cohen [52] pay particular attention to the four dimensions of suburban sprawl: residential density, land use mix, degree of centering, and street accessibility. The possible explanation is focused on effects of social capital on mental health, which depends on the close relationships like family members, partners, and friendships instead of the relations with neighbors [52]. Despite the controversial findings, there seems to be a consensus that the social aspects of sprawl pose some threats on the mental health, such as automobile commuting and the sense of social isolation and loneliness, taking a toll on mental health [18].

Thus, the relations between suburban environment and mental health are dubious even in the Western context. In China, the mental health consequences of suburban environment are far from established since there are fewer studies on this topic. However, it is suspected that the attributes traits that differentiate metropolitan places from one another could affect mental health [43].

In this sense, this study is of substantial significance in terms of the empirical study on the relationships between the suburban environment and mental health in developing countries like China.

3.3 Conclusion

This chapter has investigated empirical studies about the health implications of suburban living in China from two domains. First, focusing on the commodity housing estates, the development issues against the connotation of healthy neighborhood are addressed based on the investigation of commodity housing development in China, taking Guangzhou as a case. With the expansion to suburbs, the trend of suburban living has been revealed, which is promoted by SCHEs. Four characteristics of SCHEs are identified: urban peripheral location, leapfrogging development pattern, packaged enclaves, and middle-income population. In addition, three stages of the SCHEs development in Guangzhou are discussed: market-government cooperation, developer-dominated, and planning regulation enhanced periods. Based on this, this study concludes that the issues adverse to the requirements of health-promoting environment rely both on market-driven oriented development principles and discrepancies between market demands and health needs, especially in terms of commuting, heavy traffic, and lack of public services.

Second, two types of health threats that stemmed from suburban living have been identified based on the examination of the relationships between health and urban environment: obesity and mental disorders. Given the prevalence of obesity among urban Chinese, especially the middle-income population, the rise of SCHEs seems to encourage a sedentary lifestyle, physical inactivity and reduced energy expenditure, resulting in the increased incidence of obesity. Although most studies affirm that cities are negatively related with mental disorders (especially the depression and anxiety) and there are different incidences of mental disorders in various urban environments, there has been insufficient empirical evidence of the relations between the suburban environment and mental health. Thus, with the limited number of empirical studies conducted in Chinese suburbs, this study adds to the literature on suburban living and human health, focusing on the relations between the physical and social characteristics of SCHEs and the prevalence of obesity and mental disorders.

References

1. Bauman A, Allman-Farinelli M, Huxley R, James WPT (2008) Leisure-time physical activity alone may not be a sufficient public health approach to prevent obesity—a focus on China. Obes Rev 9(s1):119–126. https://doi.org/10.1111/j.1467-789X.2007.00452.x
2. Blair SN (2009) Physical inactivity: the biggest public health problem of the 21st century. Br J Sports Med 43(1):1–2
3. Breitung W (2012) Enclave urbanism in China: attitudes towards gated communities in Guangzhou. Urban Geogr 33(2):278–294. https://doi.org/10.2747/0272-3638.33.2.278
4. Brownson RC, Boehmer TK, Luke DA (2005) Declining rates of physical activity in the United States: what are the contributors? Annu Rev Public Health 26(1):421–443. https://doi.org/10.1146/annurev.publhealth.26.021304.144437
5. Chen CM (2008) Overview of obesity in Mainland China. Obes Rev 9(s1):14–21. https://doi.org/10.1111/j.1467-789X.2007.00433.x

6. Chen J, Guo F, Wu Y (2011) One decade of urban housing reform in China: urban housing price dynamics and the role of migration and urbanization, 1995–2005. Habitat Int 35(1):1–8. https://doi.org/10.1016/j.habitatint.2010.02.003
7. Chen Y, Zheng Z, Yi J, Yao S (2014) Associations between physical inactivity and sedentary behaviors among adolescents in 10 cities in China. BMC Public Health 14(1):744. https://doi.org/10.1186/1471-2458-14-744
8. Chiu RLH (1995) Commodification in Guangzhou's housing system: analysis and evaluation. Third World Plan Rev 17(3):295. https://doi.org/10.3828/twpr.17.3.2g206727607w1x48
9. Chung CJ, Inaba J, Koolhaas R, Leong ST (2001) Great leap forward: Harvard Design School project on the City. Taschen GmbH
10. Dalgard OS, Tambs K (1997) Urban environment and mental health: a longitudinal study. Br J Psychiatry 171(6):530–536. https://doi.org/10.1192/bjp.171.6.530
11. Davis D (2000) The consumer revolution in urban China. University of California Press
12. Ewing R, Schmid T, Killingsworth R, Zlot A, Raudenbush S (2003) Relationship between urban sprawl and physical activity, obesity, and morbidity. Am J Health Promot 18(1):47–57. https://doi.org/10.4278/0890-1171-18.1.47
13. Faris REL, Dunham HW (1939) Mental disorders in urban areas: an ecological study of schizophrenia and other psychoses. Univ. Chicago Press, pp xxxviii, 270
14. Feng J, Zhou Y, Wu F (2008) New trends of suburbanization in Beijing since 1990: from government-led to market-oriented. Reg Stud 42(1):83–99. https://doi.org/10.1080/00343400701654160
15. Fleischer F (2007) To choose a house means to choose a lifestyle. The consumption of housing and class-structuration in urban China. City & Society 19(2):287–311
16. Fleischer F (2010) Suburban Beijing: housing and consumption in contemporary China. University of Minnesota Press
17. French P, Crabbe M (2010) Fat China: how expanding waistlines are changing a nation. Anthem Press.
18. Frumkin H (2002) Urban sprawl and public health. Public Health Rep 117
19. Frumkin H, Frank LD, Jackson RJ (2004) Urban sprawl and public health: designing, planning, and building for healthy communities. Island Press
20. Gong P, Liang S, Carlton EJ, Jiang Q, Wu J, Wang L, Remais JV (2012) Urbanisation and health in China. The Lancet 379(9818):843–852. https://doi.org/10.1016/S0140-6736(11)61878-3
21. Halpern D (1995) Mental health and the built environment: more than bricks and mortar? Taylor & Francis US
22. Hare EH, Shaw GK (1965) Mental Health on a new housing estate: a comparative study of health in the two districts of Croydon (Issue 12). Oxford University Press
23. He JJ, Jia B (2007) Sustainable residential landscapes: a case study in Guangzhou, China. Landsc Res 32(2):241–254. https://doi.org/10.1080/01426390701231747
24. Hu X, Cook S, Salazar MA (2008) Internal migration and health in China. The Lancet 372(9651):1717–1719. https://doi.org/10.1016/S0140-6736(08)61360-4
25. Hua E (2001) A study on residential suburbanization in Guangzhou [Guangzhou shi zhu zhai jioa qu hua fa zhan yan jiu]. J Guangdong Univ Technol (Soc Sci Ed) 1(1):78–80
26. Hui ECM, Wong FKW (1999) Housing Reform in Guangzhou and Shenzhen, China. Rev Urb Regional Dev Stud 11(2):141–152. https://doi.org/10.1111/1467-940X.00012
27. Jim CY, Chen WY (2007) Consumption preferences and environmental externalities: a hedonic analysis of the housing market in Guangzhou. Geoforum 38(2):414–431.
28. Kerr J, Rosenberg D, Frank L (2012) The role of the built environment in healthy aging: community design, physical activity, and health among older adults. J Plan Lit 27(1):43–60. https://doi.org/10.1177/0885412211415283
29. Krabbendam L, van Os J (2005) Schizophrenia and urbanicity: a major environmental influence—conditional on genetic risk. Schizophr Bull 31(4):795–799. https://doi.org/10.1093/schbul/sbi060

30. Lawrence RJ (2002) Healthy residential environments. In: Handbook of environmental psychology, pp 394–412
31. Lederbogen F, Kirsch P, Haddad L, Streit F, Tost H, Schuch P, Wüst S, Pruessner JC, Rietschel M, Deuschle M, Meyer-Lindenberg A (2011) City living and urban upbringing affect neural social stress processing in humans. Nature 474(7352), Article 7352. https://doi.org/10.1038/nature10190
32. Lee RL, Loke AY (2011) Lifestyle behaviours and psychosocial well-being of Chinese adolescents in Hong Kong and Guangzhou, China: a cross-sectional comparative survey. J Clin Nurs 20(19–20):2733–2743. https://doi.org/10.1111/j.1365-2702.2011.03737.x
33. Li L (2005) Exploring a new dimension of residential differentiation in urban China under market transition: a study of suburban residential enclaves. Hong Kong Baptist University, Centre for China Urban and Regional Studies
34. Li L (2007) A home of one's own: Ways of life in privately governed commodity housing estates in urban China. ProQuest
35. Li P (2008) The reform and opening up of Panyu. Nanfang Daily Press
36. Li S, Zhu Y, Li L (2012) Neighborhood type, gatedness, and residential experiences in Chinese cities: a study of Guangzhou. Urban Geogr 33(2):237–255. https://doi.org/10.2747/0272-3638.33.2.237
37. Lin F (2008) The study of countermeasure about the gread residential district in Guangzhou "Huanan Project.". Master thesis, Tongji University
38. Lin L, Ou Y (2004) Analysis on the features of evolvement of residential district in Guangzhou after reform and opening up. Planners 20(9):66–70
39. Lin L, Xiao L, Chen C (2009) Residential suburbanization in Guangzhou: a perspective of residents [shi min shi jiao de Guangzhou shi ju zhu jiao qu hua yan jiu]. J South China Normal Univ 2:117–121
40. Lü J, Rowe PG, Zhang J (2001) Modern urban housing in China 1840–2000. Prestel Pub
41. Ma GS, Li YP, Wu YF, Zhai FY, Cui ZH, Hu XQ, Luan DC, Hu YH, Yang XG (2005) [The prevalence of body overweight and obesity and its changes among Chinese people during 1992 to 2002]. Zhonghua Yu Fang Yi Xue Za Zhi [Chin J Prev Med] 39(5):311
42. Martin FM, Brotherston JHF, Chave SPW (1957) Incidence of neurosis in a new housing estate. Br J Prev Soc Med 11(4):196–202
43. Oliver JE (2003) Mental life and the metropolis in Suburban America: the psychological correlates of metropolitan place characteristics. Urban Aff Rev 39(2):228–253. https://doi.org/10.1177/1078087403254445
44. Peen J, Schoevers RA, Beekman AT, Dekker J (2010) The current status of urban-rural differences in psychiatric disorders. Acta Psychiatr Scand 121(2):84–93. https://doi.org/10.1111/j.1600-0447.2009.01438.x
45. Peng M (2008) The development and planning measures of commercial housing in metropolitan suburbs and its countermeasures in urban planning—a case study of Shanghai. Planners 24(4):21–25
46. Popkin BM (2008) Will China's nutrition transition overwhelm its health care system and slow economic growth? Health Aff 27(4):1064–1076. https://doi.org/10.1377/hlthaff.27.4.1064
47. Pretty J, Peacock J, Sellens M, Griffin M (2005) The mental and physical health outcomes of green exercise. Int J Environ Health Res 15(5):319–337. https://doi.org/10.1080/09603120500155963
48. Ren X (2010) Territorial expansion and state rescaling: a critique of suburbanization studies in China. In:@@ M. Clapson & R. Hutchison (Eds.), *Suburbanization in Global Society* (Vol. 10, pp. 347–366). Emerald Group Publishing Limited. https://doi.org/10.1108/S1047-0042(2010)0000010017
49. Sarkar C, Webster C, Gallacher J (2014) Healthy cities: public health through urban planning. Edward Elgar Publishing
50. Shen J, Wu F (2013) Moving to the suburbs: demand-side driving forces of suburban growth in China. Environ Plan A Econ Space 45(8):1823–1844. https://doi.org/10.1068/a45565

51. Shen Y-C, Zhang M-Y, Huang Y-Q, He Y-L, Liu Z-R, Cheng H, Tsang A, Lee S, Kessler RC (2006) Twelve-month prevalence, severity, and unmet need for treatment of mental disorders in metropolitan China. Psychol Med 36(2):257–267. https://doi.org/10.1017/S0033291705006367
52. Sturm R, Cohen DA (2004) Suburban sprawl and physical and mental health. Public Health 118(7):488–496. https://doi.org/10.1016/j.puhe.2004.02.007
53. Sullivan WC, Chang C-Y (2011) Mental health and the built environment. In: Dannenberg AL, Frumkin H, Jackson RJ (Eds) Making healthy places: designing and building for health, well-being, and sustainability. Island Press/Center for Resource Economics, pp 106–116. https://doi.org/10.5822/978-1-61091-036-1_7
54. The WHO World Mental Health Survey Consortium (2004) Prevalence, severity, and unmet need for treatment of mental disorders in the world health organization world mental health surveys. JAMA 291(21):2581–2590. https://doi.org/10.1001/jama.291.21.2581
55. Wang C, Chen P, Zhuang J (2013) A national survey of physical activity and sedentary behavior of Chinese city children and youth using accelerometers. Res Q Exerc Sport 84(sup2):S12–S28. https://doi.org/10.1080/02701367.2013.850993
56. Wang D, Li S (2006) Socio-economic differentials and stated housing preferences in Guangzhou, China. Habitat Int 30(2):305–326. https://doi.org/10.1016/j.habitatint.2004.02.009
57. Wang J (2010) Planning and design of large-scale residential district. Archit Knowl S2:51–53
58. Wang Y (2008) Variation on theme of Chinese urban housing over the past two decades [zhong guo cheng shi zhu zha 20 nian zhu ti bian zou]. J Archit 4(1):1–4.
59. Wang YP, Murie A (1999) Commercial housing development in urban China. Urb Stud 36(9):1475–1494. https://doi.org/10.1080/0042098992881
60. Wang Z (2001) Da pan period: the revolution of real estate in China. Sichuan People's Publishing House
61. WHO (2001) The world health report 2001: mental health: new understanding, new hope
62. Wu F (2005) Rediscovering the 'gate' under market transition: from work-unit compounds to commodity housing enclaves. Hous Stud 20(2):235–254. https://doi.org/10.1080/026730303042000331754
63. Wu F (2010) Gated and packaged suburbia: packaging and branding Chinese suburban residential development. Cities 27(5):385–396. https://doi.org/10.1016/j.cities.2010.06.003
64. Wu F (2012) Suburban Beijing: housing and consumption in contemporary China, by Friederike Fleischer. China J 67:238–239. https://doi.org/10.1086/665777
65. Wu F, Xu J, Yeh AG-O (2006) Urban development in post-reform China: state, market, and space. Routledge
66. Wu Y (2006) Overweight and obesity in China. BMJ 333(7564):362–363. https://doi.org/10.1136/bmj.333.7564.362
67. Xiong Z-J, Zhu C-S, Qian X, Zhu J, Wu Z, Chen L (2013) Serum uric acid is associated with dietary and lifestyle factors in elderly women in Suburban Guangzhou in Guangdong Province of South China. J Nutr Health Aging 17(1):30–34. https://doi.org/10.1007/s12603-012-0076-2
68. Yan X, Zhou C, Deng S (2001) Development of commodity housing in Guangzhou and its surrounding areas. Acta Geogr Sin-Chin Ed 56(5):570–580
69. Yeung CW (1984) Large-scale private housing development in the suburban area in Hong Kong a case study of Fairview Park. University of Hong Kong, Centre of Urban Studies and Urban, Planning
70. Zhang J, Shi X, Liang X (2013) [Economic costs of both overweight and obesity among Chinese urban and rural residents, in 2010]. Zhonghua liu xing bing xue za zhi = Zhonghua liuxingbingxue zazhi 34(6):598–600
71. Zhang X, Shen L, Wu Y (2011) Green strategy for gaining competitive advantage in housing development: a China study. J Clean Prod 19(2):157–167. https://doi.org/10.1016/j.jclepro.2010.08.005
72. Zhou J, Luo Q (2007) Development pattern of suburban residence in the suburbanization of metropolis. Plan Stud 3:25–29

73. Zhou M, Logan JR (1996) Market transition and the commodification of housing in urban China*. Int J Urban Reg Res 20(3):400–421. https://doi.org/10.1111/j.1468-2427.1996.tb00325.x
74. Zhou Y, Logan JR (2005) Suburbanization of urban China: a conceptual framework
75. Zhou Y, Logan JR (2008) Growth on the edge: the new Chinese Metropolis. In: Urban China in transition. Wiley & Sons, Ltd., pp 140–160. https://doi.org/10.1002/9780470712870.ch6
76. Zhu J (2000) The changing mode of housing provision in transitional China. Urban Affairs Review 35(4):502–519. https://doi.org/10.1177/10780870022184507
77. Zou Y, Zhang R, Zhou B, Huang L, Chen J, Gu F, Zhang H, Fang Y, Ding G (2015) A comparison study on the prevalence of obesity and its associated factors among city, township and rural area adults in China. BMJ Open 5(7):e008417. https://doi.org/10.1136/bmjopen-2015-008417

Chapter 4
Analytical Framework and Research Method

As presented in Chap. 1, this study aims to formulate recommendations for developing healthy neighborhoods in China's suburban environments by investigating the health performance of suburban residents and the factors of the neighborhood environment influencing the health. Three research questions are proposed: *(1) What is the health performance of the residents in SCHEs of Guangzhou? (2) What factors derived from SCHEs' living environment affected the health performance of residents, and how? (3) How have the institutional factors in housing development caused the health implications of the neighborhood living environment?*

Building upon the theoretical foundation laid in Chap. 2 and the empirical context presented in Chap. 3, this chapter presents the analytical framework, the case study and the data collection methods, to answer the three research questions. This chapter consists of two parts. The first part is a three-level analytical framework in line with the three research questions, including the evaluation of the health performance, the analysis of determinants embedded in the residential environment and residents' behaviors, and the explanation for influential mechanisms. The second part focuses on the selection of research methods, overall research strategy, the case selection strategy, data collection, and data analysis methods.

4.1 Analytical Framework

As presented in Fig. 4.1, in line with the research questions, the analytical framework consists of three levels. The first level examines whether or not suburban neighborhoods are healthy, using three SCHEs developed in different time periods as case studies. The health performance of residents in SCHEs is evaluated in terms of physical and mental health. In addition, the hypothesis that the residents' health of suburban neighborhoods built in different time periods is significantly different is tested.

T. Zhang, *Towards Healthy Settlements*, Urban Sustainability,
https://doi.org/10.1007/978-981-97-1207-6_4

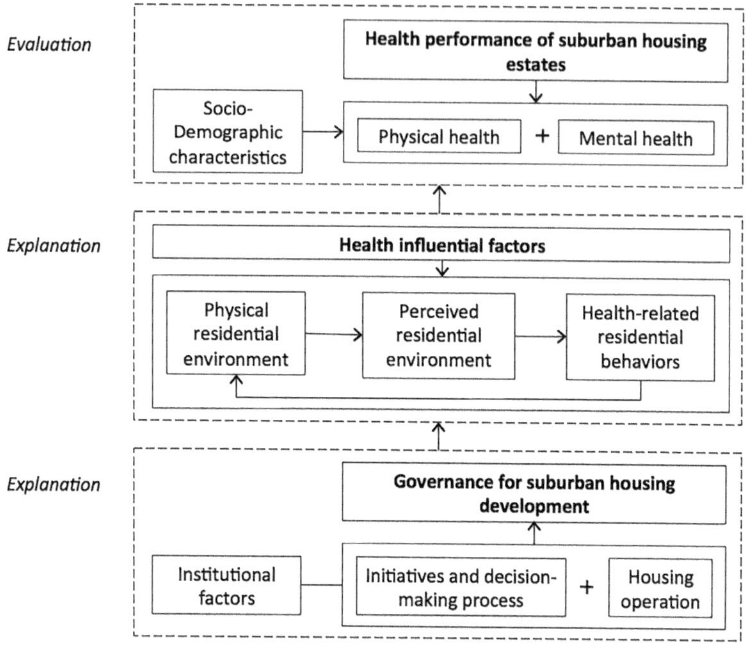

Fig. 4.1 Analytical framework

The second level of analysis has two sub-stages: the first sub-stage identifies the factors that affect residents' health based on the investigation of the suburban living environment. As discussed in Chap. 2, residential environmental factors and health-related behavioral factors are investigated, and the associations between behavioral factors and residents' health, and between environmental factors and residents' health are analyzed to identify the factors influencing health. Thus, the puzzle that whether the SCHEs have provided health-promoting residential environment and promoted health-promoting residential behaviors could be solved. In addition, adopting the human ecological perspective, it is proposed that both residential environment and health-related residential behaviors could determine the health performance, while it is the mode of people-environment transaction that have more fundamental effects on residents' health. The second sub-stage focuses on the interactions between environmental and behavioral factors that influence residents' health. An ecological principle of reciprocal people-environment transactions is adopted in the analysis. Three pairs of relations are elaborated: physical and perceived residential environment, perceived residential environment and health-related residential behaviors, and the effects of residential behaviors on the physical residential environment (Fig. 4.1).

The third part explains the institutional factors of housing development that affect the formation of residential environment. An effort is made to identify the institutional factors in the two phases of housing development: initiatives and decision-making of suburban housing development and housing management. For the development phase, the roles of three key players are analyzed: the local government, the developer, and the planners. Interplays among these actors have significant impacts on the formation of the physical plans, which determines physical residential environment, affecting the factors that influence health. In the housing management phase, the interactions among the housing manager, residents' committee and residents in terms of housing estate affairs are analyzed (Fig. 4.1). The detailed analytical framework is in the next section.

4.1.1 *Evaluation of the Health Performance*

The first level of analysis is to judge whether the SCHEs in Guangzhou are healthy and to what extent. This could be operationalized by examining the health performance of residents. Initially, the unit of analysis should be clarified. As discussed in Chap. 2, healthy neighborhood is conceptualized as a harmonious people-environment relationship, promoting physical health, mental health, and social well-being. This conceptualization indicates that health should be captured at both the individual and collective levels, however, in this study, the unit of analysis is set of the individual within her or his environment [34]. The evaluation of the health performance of residents in SCHEs focuses on physical and mental health (Fig. 4.3).

To assess physical health, two indicators are used based on the literature review and time and resources availability. First, the literature reports that general physical health is often applied by scholars in social science, and it is measured with self-assessed health or by accessing functional ability or perceived health with self-assessed health or functional ability or perceived health [33]. So it is rational to use self-reported general physical health to indicate the physical health of residents as presented in Table 4.1. Second, referring to Chap. 2, in providing policy recommendation for health promotion, the extant literature focuses on the relations of built environment with obesity or overweight. At the same time, the empirical background of suburban living in Chap. 3 also suggests that the SCHEs could shape to a sedentary lifestyle, increasing the risks of being overweight and obese. Thus, the second indicator for assessing the physical health is the prevalence of obesity, as measured by the body mass index (BMI).

For mental health, two indicators are selected as presented in Table 4.2. According to the symptoms of mental disorders in epidemiologic studies reviewed in Chap. 2 and mental health condition of Chinese residents demonstrated in Chap. 3, three symptoms of mental disorders are selected in this study: depression, anxiety and stress. In addition, two levels of mental disorders (everyday psychopathological and clinical disorder) are differentiated. For depression, the clinical diagnosis is adopted based on the CES-D self-report depression scale, which is the widely-used scale for depression measure [16] (Table 4.2). This scale is weighted by the frequency of occurrence during the past week, and there are 20 items selected from a pool of items based on clinical literature and factor analytical studies [22]. The reliability and validity in the general population have been proved, with scores highly correlated

Table 4.1 Evaluation indicators of health performance of residents

Domain	Elements	Indicators	Measurement
Health performance	• Physical health	• Self-report general physical health condition • Obesity	• Ordinal data • Binary data
	• Mental health	• Depression disorder[a] • Risk of mental disorders (level of anxiety; level of stress; effects of anxiety/stress on residential living)	• Binary data • Continuous data

[a] The level of depression was measured with the CES-D scale a self-report depression scale during the questionnaire survey. More than half respondents (58.6%) refused to complete the whole scale, so they were not subjected to analysis

with diagnoses of clinical depression [22, 23, 25]. With the translated version from standard English version, the CES-D scale has been validated in Chinese population previously, and several scholars focusing on depression symptoms of Chinese population have adopted this scale accordingly [15].

The second indicator is the risk of mental disorders (RMD), measuring mild symptoms of anxiety and stress rather than clinical disorders. This choice is based on several considerations. The prevalence of mental disorders in China is quite low, perhaps because of stigma-related under-reporting bias [24]. In addition, negative emotions are important health risks in terms of all-cause mortality [30]. Thus, this study develops a three-item five-point scale to collect the information of anxiety and stress. Both are weighted by the frequency of occurrence in the past month, and the impact of these two mild symptoms on daily residential living is the third item. For the measurement, responding to the critiques of dichotomy justification of cases and non-cases in Chap. 2, a continuous measurement is generated based on the three items with the same weight (Table 4.1).

Furthermore, the health performance of the residents is compared across the three suburban neighborhoods, which represent the three types of SCHEs built in three different time periods (Fig. 4.3). Based on the comparison analysis, the hypothesis that suburban neighborhood type has influential effect on health performance could be tested. As pointed out in Chap. 2, health outcomes vary across populations with different demographic and socioeconomic status. Demographic characteristics such as age, gender, income, and education could intervene between the objective world and an individual's evaluation of it, and individual experience, the aspiration level or expectations of individual, and individual's cultural background [20]. Thus, the socio-demographic characteristics considered in this study involve age, gender, income, education status, employment status, family structure, residence length, and homeownership (Fig. 4.3). In addition, these factors should be controlled in the next step of analysis.

Table 4.2 Scoring for center for epidemiological studies depression scale (CES-D)

During the past week…	Rarely or none of the time (less than 1 day)	Some or a little of the time (1–2 days)	Occasionally or a moderate amount of time (3–4 days)	All of the time (5–7 days)	Score
1. I was bothered by things that usually don't bother me	0	1	2	3	
2. I did not feel like eating; my appetite was poor	0	1	2	3	
3. I felt that I could not shake off the blues even with health from my family	0	1	2	3	
4. I felt that I was just as good as other people	3	2	1	0	
5. I had trouble keeping my mind on what I was doing	0	1	2	3	
6. I felt depressed	0	1	2	3	
7. I felt that everything I did was an effort	0	1	2	3	
8. I felt hopeful about the future	3	2	1	0	
9. I thought my life had been a failure	0	1	2	3	
10. I felt fearful	0	1	2	3	
11. My sleep was restless	0	1	2	3	
12. I was happy					
13. I talked less than usual	0	1	2	3	
14. I felt lonely	0	1	2	3	
15. People were unfriendly	0	1	2	3	
16. I enjoyed life	3	2	1	0	
17. I had crying spells	0	1	2	3	
18. I felt sad	0	1	2	3	
19. I felt that people dislike me	0	1	2	3	
20. I could not "get going"	0	1	2	3	
Total score					

Note Total score of 16 or higher is considered depressed
Source Radloff [22]

4.1.2 Exploration of Factors Influencing Health

For health determinants, environmental and behavioral factors are highlighted in Chap. 2, showing that the determinants for health performance of SCHEs could be explored in terms of residential environment and health-related residential behavior. From the human ecological perspective, the relationships between the residential environment and health-related residential behavior are essential for achieving the healthy neighborhood, that is, the residents-environment transactions could have great health implications in suburban living. The analysis framework for exploring the health determinants is shown in Fig. 4.2.

4.1.2.1 Residential Environment

In studying the factors that affect health, the residential environment should be investigated. An operational definition of residential environment should be developed initially. With the professional training background of architecture and urban design, my foci of the physical environment mainly include the manmade physical components of human settlements [12]. In addition, based on the literature and observation, physical components of a residential neighborhood include (1) housing units, (2) residential amenities; (3) transportation system, including the streets network within residential neighborhood and the transport facilities connecting with the urban transport system; (4) infrastructure; (5) green and open space; (6) ambient environment [2–4, 11, 17, 19, 32]. These elements could be categorized for further analysis. Based on Chap. 2, this study groups the elements of residential environment into four domains: (1) transportation environment, including public transportation system,

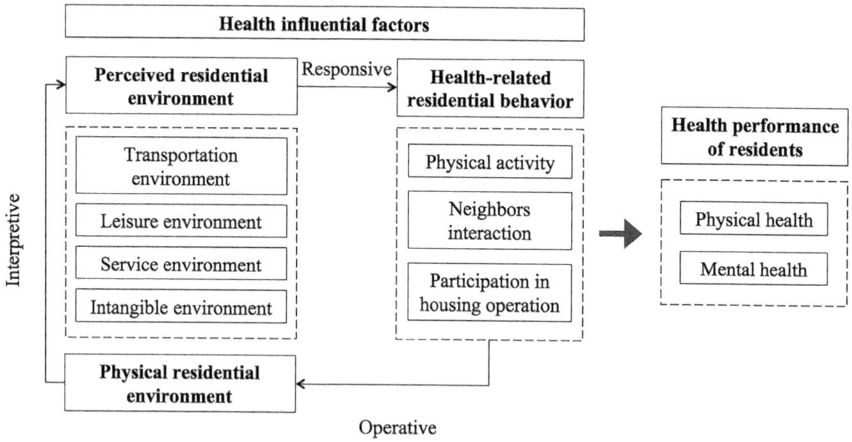

Fig. 4.2 Health influential factors and health outcomes

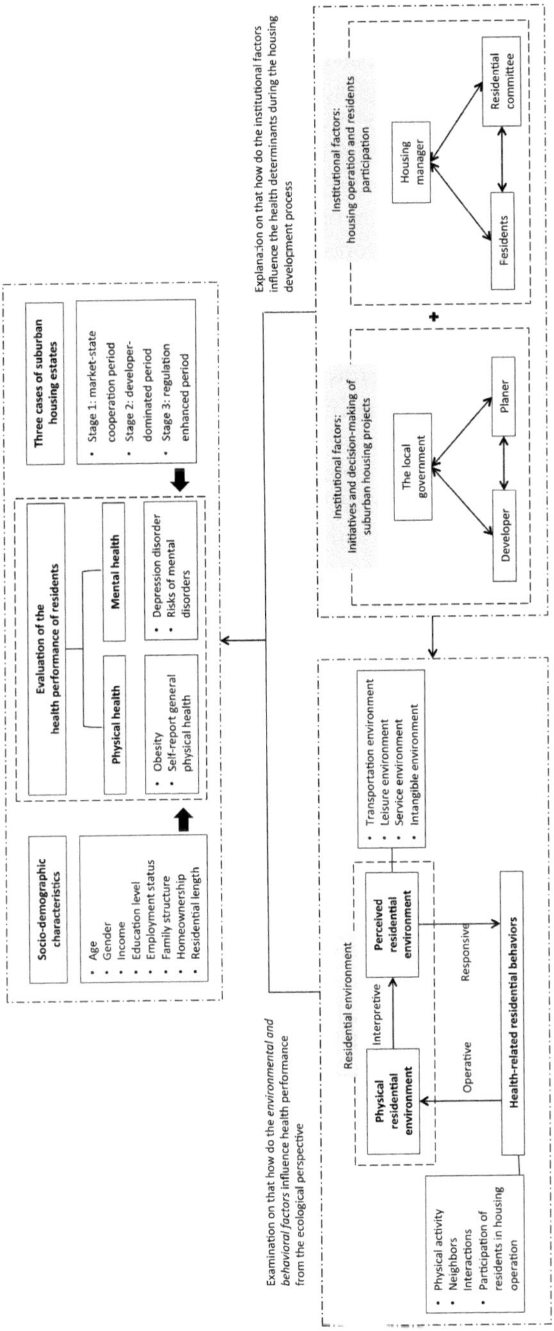

Fig. 4.3 Analytical flow of the three analysis steps

street network and walking environment within the neighborhood; (2) leisure environment, including facilities and space provided by the neighborhood for physical leisure activities; (3) service environment, especially refers to the educational, health care and daily commercial services; and (4) design features, comprised by density, safety, greenness, symbolic meaning and aesthetic qualities.

From the human ecological perspective, influential factors in residential environment in this study are analyzed in terms of the physical residential environment and the perceived residential environment (Fig. 4.2). Theoretically, both subjective and objective evaluations are necessary in order to understand the quality of the urban environment [7, 20], and especially for illustrating the people-environment relationship [2, 7, 17, 20]. In people-environment transactions, "each person has a behavioral environment which is the environment as perceived" [34, p. 21], and the perception proceeds is innate abilities of human, through which they organize environment stimuli into coherently structured forms or patterns, determining human behaviors (Walmsley and Lewis [34], as cited in Kirk 1963). For the same environment stimuli, although each individual could perceive it in a unique way, the resultant behavioral environments have much in common because "they are derived from both common neurological mechanisms that are innate in people and from common superimposed socializing experiences" (Walmsley and Lewis [34, p. 21], as cited in Ittelson et al. 1974).

Accordingly, to measure the residential environment, objective and subjective indicators are adopted. Though there is a continuing tension between subjective and objective use of the indicators (Turcu [31], as cited in Astleithner and Hamedinger 2003 and Rydin et al. 2003), the evaluation of objective and subjective considerations is supported by the operationalization of the evaluation indicators by many scholars [2, 7, 17, 20]. Objective indicators emphasize the ease for measuring and understanding and reflect the local circumstance, subjective indicators indicate the inherent result of the subjective indicator selection process [31]. Objective indicators could be from the experts' technical evaluation of the same place or from the investigation of the communities, while subjective indicators come from residents' perceived quality of the neighborhood (perception and satisfaction) [2]. Table 4.3 lists the objective indicators and perceived indicators for measuring the residential environment.

Furthermore, four criteria for assessing health-promoting capacity of residential environment are developed based on the literature review in Chap. 2: (1) promoting active travel; (2) increasing opportunities for physical exercises; (3) providing convenient and adequate service facilities; and (4) reducing environment stress and creating positive design features (Table 4.3). Accordingly, key elements of the four aspects of residential environment are selected as the potential influential factors for residents' health, which will be evaluated to explore the significant health determinants. (Table 4.3).

Table 4.3 Framework for health determinants derived from residential environment

Domain	Criteria	Elements	Indicators	
			observed (objective)	Perceived (subjective)
Residential environment	• Promoting active travel	• Transportation system • Destination (facilities) • Walking environment	• Street network pattern • Public transportation • Accessibility of service facilities • Quality and characteristics of streets	• Perception of accessibility of destinations by walking • Perception of the quality of walking surface
	• Increasing opportunities for physical exercise	• Fitness facilities • Open space	• Quantity and accessibility of fitness facilities • Quantity and accessibility of open space	• Place preference for physical exercises • Perception of the quantity of fitness facilities • Perception of the quantity of and open space • Satisfaction with the quality of space and facilities
	• Providing convenient and enough service facilities	• Health service facilities • Daily commercial facilities • Education facilities	• Quantity and accessibility of health service facilities • Quantity and accessibility of daily service facilities • Quantity and accessibility of primary and middle schools	• Places chosen for health issues • Satisfaction with the schools, medical care and daily service facilities

(continued)

Table 4.3 (continued)

Domain	Criteria	Elements	Indicators	
			observed (objective)	Perceived (subjective)
	• Reducing environmental stress and creating positive design feature	• Safety • Density • Green and open space • Ambient environment	• Safety management • Population/ building density • Green ratio • Noise and cleanness	• Perception of safety • Perceived density • Perception of green space • Perception of ambient environment
		• Symbolic meaning • Aesthetic qualities	• Design ideology • Attributes of public realm, landscape	• Perception of public realm • Sense of aesthetic • Sense of belonging

4.1.2.2 Health-Related Residential Behaviors

Many studies have noted the links between the behavioral factors and health. Health-related behaviors mainly comprise physical activity, dietary behavior, health screening, and collective behavior that generate social capital. In the context of suburban living, as indicated in Fig. 4.2, this study focuses on three types of health-related behaviors in residential life: physical activity (travel activity and leisure activity), interactions among neighbors and residents' participation in housing estate operation.

First, physical activity is an important behavioral factor affecting human health and wellbeing. Two types of physical activity are examined: (1) *transportation activity* (utilitarian), undertaken to accomplish another purpose, such as walking and cycling to walk, school, or the store; and (2) *leisure activity*, undertaken for discretionary reasons on leisure time, such as jogging, weight lifting, hiking, basketball, and soccer [8]. The mode, duration, and frequency of the physical activity are crucial in terms of their influence on health outcomes (Table 4.4).

Second, studies have pointed out that interactions among neighbors are essential for the social well-being of the neighborhood, which in turn has crucial associations with mental health. In this context, this study uses interaction among neighbors as a health-related residential behavior that reflects the social cohesion of the neighborhoods. In the operation, two categories of neighbors' interactions are identified: formal (organized interactions by housing manager or residents' committee) and informal or unplanned interactions (such as conversations with neighbors, helping each other, playing cards, casual meetings, or seeing a regular at the coffee shop),

Table 4.4 Framework for health determinants derived from health-related residential behaviors

Domain	Criteria	Elements	Indicators	
			Observed (objective)	**Perceived (subjective)**
Health-related residential behaviors	• Nurturing regular physical activity (meet recommendations)	• Travel activity[a] • Leisure activity	• Travel type • Walking time • Commuting time • Variety of leisure activities • Intensity of leisure activities	• Willingness to walk • Preference for leisure activity • Regularity of doing leisure activities
	• Promoting neighbors interactions	• Informal interactions • Formal interactions	• Variety and frequency of informal interaction • Variety and frequency of community activities	• Willingness to attend informal interactions • Willingness to attend formal community activities • Perception of neighbors interaction
	• Promoting residents' participation in housing operation	• Residents' participation and satisfaction	• Variety and frequency of participating housing operation	• Concerns on neighborhood affairs • Satisfaction with housing operation affairs

[a] McGinn AP, Evenson KR, Herring AH, Huston SL, Rodriguez DA (2007) Exploring associations between physical activity and perceived and objective measures of the built environment. J Urban Health 84(2):162–184

which are equally important in generating social capital [6]. In terms of measurement, both objective (behavior) and subjective (perception or opinions) indicators are developed to examine the interaction (Table 4.4).

Third, the participation of residents in managing the housing estate could also impact on their health outcomes. So the residents' participation in the housing management is investigated here as one type of health-related residential behavior, as listed in Table 4.4.

For assessment, three criteria for assessing health-related residential behaviors are identified: (1) regular physical activity; (2) promoting neighbors' interactions; (3) satisfaction with housing management and development. Consistent with the objective and subjective evaluation methods, both objective and subjective indicators for health-related residential behavior are developed in Table 4.4. The objective indicators are used to explore the actual conducted residential behaviors such as the type, intensity, frequency, and duration of the behaviors, while the subjective ones are used to measure the perceptions of residents such as the willingness, preference and satisfaction.

4.1.3 Explanations for the Factors Influencing Health

The third part of analysis describes the factors that affect health in relation to the human ecological perspective and institutional factors in the property development process.

4.1.3.1 Residents-Environment Transactions

To ascertain whether it is the mode of people-environment transactions that affect the health of the residents, it is necessary to examine the effects of the residents-environment transaction mode on the factors influencing health, keeping in mind that the people-environment congruity is an important predictor of health and well-being [27, 28]. The theoretical foundation for analyzing the residents-environment transactions is mainly inspired by the work of Stokols. The people-environment congruity considers the match between individual life satisfaction and objective standards of living and the extent of the possibilities for people to fulfill their goals and needs through their environment [18]. Theoretically, two assumptions form the foundation for understanding the residents-environment transactions. The first is human–environment optimization, assuming that "people ideally strive to achieve 'optimal environments', or those that maximize the fulfillment of their needs and the accomplishment of their goals and plans" [29, p. 30]. The second is that cognition and perception are the bridges between the objective environment and the reactions of human. In other words, cognitive and psychological factors mediate the effects of environment on people [29]. So the assessment of residents environment transactions should highlight the identification of the two pathways: whether they can only passively adapt to their residential environment, whether the residents could improve or change their residential environment.

Thus, three study objects in term of suburban residential living are identified to investigate the residents-environment transactions: physical residential environment, perceived residential environment, and health-related residential behaviors. Referring to the four modes of people-environment transactions demonstrated in Chap. 2 (interpretive, evaluative, operative, responsive), three modes could be identified among the above three objects: interpretive (physical residential environment—perceived residential environment), responsive (perceived residential environment—health-related residential behaviors) and operative (health-related residential behaviors-objective residential enjoinment) (Fig. 4.2). The interpretive mode could be used to explain the factors influencing health derived from the relations between physical and perceived residential environment. Through their cognition and experience, residents interpret the physical environment of the neighborhood to form the perceived residential environment. The process of interpretation is in fact a mode of people environment transaction. In addition, the responsive mode could be explained as adaption and reaction of residents to their perceived residential environment. Specifically, the pattern of the physical activity is determined by opportunities and possibilities provided by the residential environment, and the pattern of neighbors' interactions is either promoted or constrained by the opportunities provided by the residential environment. Moreover, considering optimization, the operative mode is expected to appear in the relations between health-related residential behavior and physical residential environment. It is proposed that residents seek ways to modify their surroundings to fulfill their living demand, and residents' participation in the modification process through participating in the management of the housing estate is examined to find out whether the operative mode is effective in maintaining a desirable environment. On the whole, it hypothesizes that only when the three modes are all effective among the three objects and formulate a complete circuit as shown Figs. 4.2 and 4.3, that the neighborhood could be regarded as a health-promoting environment.

4.1.3.2 Institutional Factors in Housing Development Process

It is hypothesized that it is the institutional factors in the suburban housing development could determine the factors influencing health. The roles of different stakeholders in suburban housing development should be investigated, as well as their interplays.

Stakeholders in housing development should be identified as a first step to understand the institutional factors. Referring to Bhatti's housing life-cycle framework, the phases of housing development range from inception, design, construction, building use, to refurbishment and demolition [5]. Based on this, the analyses on the institutional factors affecting factors influencing health focus on two stages: initiatives and decision-making stage and housing management (Fig. 4.3), during which the key stakeholders involved could be categorized into five groups: government, developer, planner, resident, and housing managers [5]. The first three groups play more roles in the phases of inception, planning, design and construction, which transform natural resources into livable spaces, the last two groups play more roles in the phase of housing management [1, 5]. Accordingly, government, developer, and planner could manipulate environmental factors during residential development, indirectly influencing the quality of residential environment. Furthermore, the management of housing conducted by both housing manager and residents, and committee could influence the quality of the residential environment, and residents' participation in housing management is also considered influential (Fig. 4.3).

Additionally, the roles of and interactions across stakeholders are influential in the formation of the residential environment, which affect the health of residents accordingly. The local governments are significant in the development of China's suburbs, organizing or orchestrating the territorialization of the municipal state in its city-region [35]. At the same time, there are both cooperation and contradictions between the municipal government and suburban district government in the growth of the suburban economy. In general, "the relationship between central city and suburban districts shifts from a top down model to a growth regime coordinated through governance" [35, p. 309]. Thus, the initiatives of municipal government and suburban district government are examined at the inception of suburban housing development, which could help to understand the fundamental determinants of the residential environment (Table 4.5).

In the market, real estate developers have a leading role in suburban housing development, motivated by the demand of capital accumulation. The orientations and strategies initiated by developers determine the characteristics of the residential environment greatly. Thus, orientation and strategies of developers for suburban housing are reflected in the decision-making process, whose effects on the residential environment are analyzed (Table 4.5). In addition, planners are responsible for making detailed plans using professional techniques, such as locating origins and destinations within space, determining the ease and type of access between them, formulating the characteristics of different spaces, and affecting the desirability of being in those spaces [8]. So the impacts of the planning and design principles adopted by planners should be considered in the formation of suburban residential environment.

Based on the examination of the roles of the three key players, their interactions should be highlighted in terms of the institutional structure of the governance mode of suburban housing development (Fig. 4.3). Three theoretical propositions are to be tested: (1) initiatives of the governments in suburban development are promoted by economic devolution and economic growth, during which the initiative of governments is irrelevant or against to that health as a public interests; (2) orientation and strategies initiated by developers lead to profit-led development, which go against to health benefits of the users partially while unintentionally facilitating the health-promoting physical elements; and (3) the decision power of the planners is limited since they have to satisfy the requirements of both the developer and the local government.

Table 4.5 Underlying factors in suburban housing development

Process	Actors	Elements to analysis
Initiative and decision making	Local government	• Initiatives of suburban development
	Developer	• Development orientation and strategies
	The planners	• Planning and design techniques
Housing operation	Housing manager	• Maintenance of the physical environment • Organization of community activities
	Residential committee	• Management of the neighborhood
	Residents	• Participation in housing operation

In addition, the institutional environment of housing management is examined, especially the new private governance with the representation of gated living space in suburbs. Under the premise that state control is pervasive and the independent self-governance by civil society is weak in the gated living space [35], the roles of residents' committee, housing manager, and residents should be analyzed (Fig. 4.3). In housing management, developer-led property management replace some functions of state provision, such as (1) the provision and maintenance of man-made facilities such as the green and open space and service facilities; and (2) the organization of the community activities, which could promote the interactions among residents and strengthen social cohesions and the social network (Table 4.5). The residents' committee, as an executive branch of local government, acts as a manager in the residential environment. In theory, the residents could optimize their residential environment by participating in housing management affairs, highlighting the importance of residents' participation in the creation of a healthy neighborhood. In this way, the interplays among these three actors are proposed to affect factors influencing health, which are analyzed with reference to the operative transaction mode (Fig. 4.3).

Finally, based on the examination on the governance mode of suburban housing development in Guangzhou, the implications and issues for achieving a health-promoting environment could be ascertained. The hypothesis to be tested is that the governance mode of suburban housing development affects the health of the residents, and the function of civil society is very weak, indicating that there is a far way towards good governance for health in the suburban housing development in Guangzhou.

4.2 Research Methods

4.2.1 Study Area and Case Selection

As demonstrated in Chap. 3, Panyu district has had a crucial role in the residential suburbanization of Guangzhou, which is a suitable study area. Located in the south periphery of Guangzhou, Panyu District has received a large number of migrations caused by urbanization and urban expansion for decades (Fig. 4.4). So Panyu is seen as the starting point for Guangzhou developing into a big city [13]. As a famous hometown of overseas Chinese (especially those living in Hong Kong and Macau), Panyu attracts lots of investment in urban construction, including housing development [14]. Commodity housing development in Panyu is typical in Guangzhou, with the typical form of multi-level residential suburban agglomeration [10].

Furthermore, housing development in Panyu could be considered a typical example with regard to the three stages of residential suburbanization demonstrated in Chap. 3. The three periods of housing development in Panyu are presented in the Fig. 4.5. Housing development in Panyu has been motivated by the capitals from

Fig. 4.4 Location of Panyu district

Hong Kong and the support of local government. Two clusters of housing estates could be identified at the northern periphery of Panyu near the Pearl River and at the north of the old central area of Panyu (Fig. 4.5). At the turn of the century, promoted by reforms and administrative adjustment, real estate development boomed in Panyu, forming residential agglomeration in the northern of the district along with the main road (Fig. 4.5). In this context, represented by the 'Eight King Kong"[1] in the north of Panyu, housing development in Panyu opened the era of large-scale housing estates

[1] Referring to the eight super large, packaged and gated communities developed from the late 1990s, lasting one decades. Including: Country Garden (*Huanan Biguiyuan*), Huanan New Town (*Huanan xincheng*), Agile Garden (*Yajule Huayuan*), Guangdi Garden (*Guangdi Huayuan*), Clifford Estate

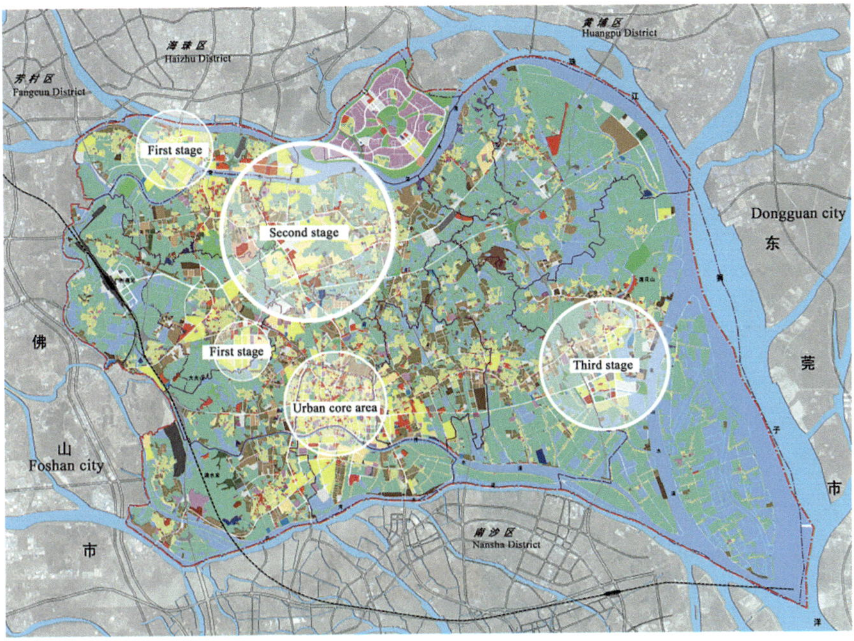

Fig. 4.5 Land use map of Panyu district *Source* Urban and rural development plan of Panyu District, 2013–2030

in China. With Panyu planned to be a new urban center of the metropolitan area according to the latest urban–rural development plan of Guangzhou, housing development has moved to the eastern and southern parts of Panyu (Fig. 4.5). Specifically, the 16th Asian Games stimulated a new round of housing development in Panyu, a new town was planned and developed on the site.

Based on the foregoing, the following criteria are considered in terms of case selection. First, considering the connotation of suburbs, peripheral location has been considered in the selection of study area of Panyu District in Guangzhou. In addition, the time factor is significant in understanding suburban neighborhoods since there are three periods of residential suburbanization in Guangzhou. Considering that suburban housing in different periods tends to have their own characteristics, neighborhoods from different stages should be chosen in the case selection. The features of the residential environment that have significant impacts on health are scrutinized for the case selection. Consistent with the demonstration in Sect. 4.1.1, four aspects of residential environment are examined in the case selection: transportation environment, leisure environment, service environment, and design features.

As a result, three neighborhoods are proposed as the case studies (Fig. 4.6). The first one is Luoxi New Town (LNT), developed from the late 1980s to 1999 under

(*Qifu xincun*), Star River (*Xinhewan*), Global Villa (*Jinxiu xiangjiang*), Nanguo Olympic Garden (*Nanguo aolinpike huayuan*).

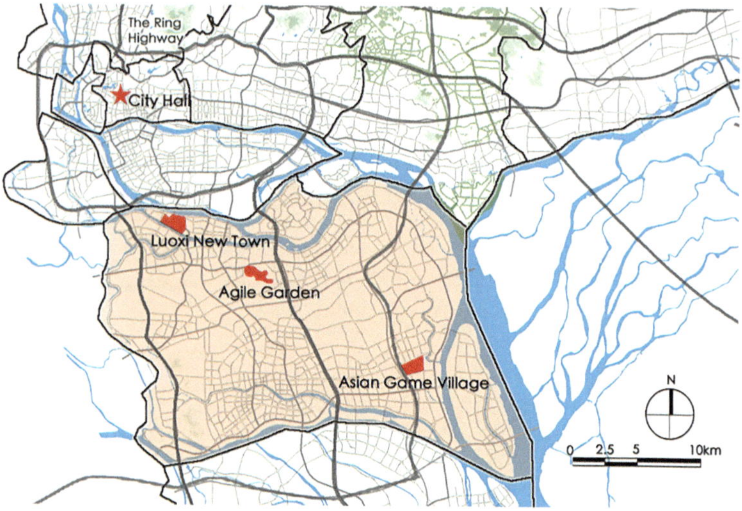

Fig. 4.6 Location of the three case neighbourhoods

the stimulation of the construction of Louxi Bridge, indicating the start of residential suburbanization in Guangzhou and representing the housing estates in the first stage of suburban housing development. The second case is Agile Garden (AG), located at the northern part of Panyu and under construction from 2000 to 2010 in the form of developer-dominated development, representing the second stage of suburban housing development. To represent the third stage of suburban housing development, Asian Game Village (GV) is selected, which was developed from 2007 to 2010 with strong government intervention and tightened planning regulation.

4.2.2 Data Collection Methods

Both qualitative and quantitative data, and objective and subjective data are collected. To answer the three research questions, three data sets are collected: (1) quantitative data for evaluating the health performance of residents; (2) quantitative and qualitative data for evaluating residential environment and health-related residential behaviors; (3) qualitative data for evaluating the governance mode of suburban housing development. Quantitative data are collected through a questionnaire survey, and qualitative data through interviews and observation.

4.2.2.1 Questionnaire Survey

Three tasks were accomplished through the questionnaire survey: (1) ascertaining the health performance of the residents; (2) asking residents' perceptions of the residential environment, (3) discovering the pattern of health-related residential behaviors. The questionnaire survey was conducted from June 15 to July 5, 2015 in the three SCHEs. It was mainly conducted in the way of self-completion, while for the elderly the face-to-face interview survey was used to help them understand the questions properly.

Sampling design

A structured questionnaire survey was undertaken to collect data set 1 and data set 2. The targeted population for the sampling in this questionnaire survey is adult residents (20 years old, included) living in the SCHEs longer than six months, comprising sampling frame[2] of the questionnaire survey. In terms of sampling design, this study adopts the modified convenience sampling rather than probability sampling for the following reasons. Frist, the three SCHEs were selected purposively, so the generalizability is not an issue. Second, there were not enough sources to conduct random sampling: the selected housing estates are gated communities and the housing managers prohibited door-to-door visits, so not all residents had the same chance of being selected for the sample. Therefore, convenience sampling is appropriate and practical in this study. Operationally, most respondents were selected in the public space of the neighborhoods randomly, such as the public squares, parks, streets, commercial shops and bus stops.

Three strategies are applied to compensate for the weakness of convenience sampling and to increase the randomness of the selected samples. Questionnaire survey was carried out at different public places in the housing estates, such as the entrance, green and open space in and around the service facilities, and bus stations. To balance the responses of the residents who do and do not work outsides, face-to-face interviews were conducted from 9:00 am to 12:00 pm and from 3:00 pm to 8:00 pm on both weekdays and weekends. Finally, the age structure of the residents was referred to in the sampling process, shown as the fraction of the targeted residents in Table 4.6.

Furthermore, the boundaries of sampling area are identified according to the jurisdiction area of one residents' committee. The boundaries of jurisdiction areas of AG and GV are easily identified by observation, while that of LNT is based on the interview of Luoxi residents' committee. The sampling area of the three cases is shown in Fig. 4.7.

Sample size is determined by several interrelated factors: sampling design, population heterogeneity, desired precision, available resources, and number of breakdowns planned in data analysis [26]. Generally, for assuring the accuracy, the desirable sample size is determined by the expected variation in the data: more varied data need

[2] Sampling frame is the operational definition of the population.

Table 4.6 Population percentage by age

Neighborhood	Age	Proportion (%)
LNT	20–29	20.10
	30–39	23.89
	40–49	17.95
	50–59	8.55
	> 60	8.05
AG	20–29	20.11
	30–39	33.54
	40–49	11.36
	50–59	7.60
	> 60	8.63
GV	20–29	21.10
	30–39	19.11
	40–49	16.97
	50–59	10.57
	> 60	12.11

Source Sixth Census of Panyu District

larger sample size[3] Although these factors are derived from simple random sampling, they also apply to nonprobability sampling designs in both a logical and intuitive sense [26]. In this study, the three cases are all the neighborhoods for suburban living of new middle-class, sharing common socioeconomic attributes, which leads to relative low level of expected variation of the data. Thus, in consideration of resources limitation and desired precision, it is acceptable to set the sample size as 1% of the number of the residents aging from 20-year-old (included) to 100-year-old in each case. Based on secondary data from official website of the local government and firsthand data from interviews, it is estimated that the total number of residents in LNT, AG, and GV are 22,230,[4] 29,520,[5] and 17,421,[6] respectively. In addition, residents who are younger than 20 years old should be excluded from this survey. Based on the proportion of the residents aged from 0–19 calculated from the census, the

[3] http://fac.arch.hku.hk/upad/apnhr/wplogin.php?redirect_to=%3A%2F%2Ffac.arch.hku.hk%2Fupad%2Fapnhr%2Fwpadmin%2F&reauth.

[4] Source: http://luopu.panyu.gov.cn/.

[5] Source: http://www.nancun.gov.cn/ShowInfo.aspx?id=20120704102500265554943000625000.

[6] The number of residents is calculated based on the interview with housing managers in GV, please refer to the Appendix 1.

Sampling area of LNT

Sampling area of AG

Sampling area of GV

Fig. 4.7 Boundaries of sampling areas of the three cases

Table 4.7 Sampling size in three cases

	Total	LNT	AG	GV
Number of residents ($\geq$ 20 years old)	54,348	17,459	23,982	13,918
Number of valid respondents	527	174	208	145
Number of respondents (sampled)	670	240	260	170
Valid response (%)	78.66	72.50	80.00	85.29
Sample size (%)	0.97	1.00	0.87	1.04

numbers of targeted residents of the three are 17,459,[7] 23,982,[8] 13,912,[9] respectively (Table 4.6). As a result, 670 questionnaires were sent out and 527 valid ones were returned (LNT, 174; AG, 208; GV, 145) (Table 4.7).

Questionnaire design

The questionnaire consists of four parts consistent with the analytical framework: (1) self-reported health conditions; (2) perceived residential environment; (3) health-related residential activities; and (4) socio-demographic status. Two types of questions were asked in the questionnaire: factual questions focusing on health-related residential activities and socio-demographic features, and attitude questions focusing on residents' perception of the residential environment and self-reported health.

For the factual questions, single choice is the main way of obtaining information. In addition, open questions are set before close-end questions when investigating the health-related residential activities, with the purpose of identifying all possible physical activities and community activities that could be conducted by residents. For socio-demographic information such as age, gender, income and height, respondents are asked to write down actual numbers. For attitude questions, both single- and multiple-choices are used. In particular, with the given statements of evaluation indicators, the five-point Likert scales are applied to elicit residents' attitudes towards the residential environment and the preference of residential behaviors. Five categories are employed for each statement (strongly agree, agree, neither agree nor disagree, disagree and strongly disagree), all of which are assigned scores. The questionnaire in found in Appendix II.

In addition, most data are collected through the self-report method, having both advantages and limitations. The advantages include "easy interpretability, richness of information, motivation to report, causal force, and sheer practicality" [21, p. 227]. Self-report is convenient and efficient in collecting data through questionnaires, and it is especially valid for measuring residents' perception of residential environment. To the contrary, self-reports suffer from issues such as "anchoring effects, credibility

---

[7] = 22230*(1–21.46%). According to the Sixth Census of Panyu District, 21.46% is the percentage of the residents younger than 20 among the total residents in LNT.

[8] = 29520*(1–18.76%). According to the Sixth Census of Panyu District, 18.76% is the percentage of the residents younger than 20 among the total residents in AG.

[9] = 17421*[1–20.14%]. According to the Sixth Census of Panyu District, 20.14% is the percentage of the residents younger than 20 among the total residents in GV.

issue, consistency seeking, self-enhancement, and self-presentation" [21, p. 228]. Respondents may have different understanding towards the same question, and their self-reports could be affected by various sources of inaccuracy [21]. Limited by the disadvantages of self-report method, questions are designed to avoid the leading questions and in the form of closed questions. In addition, questionnaire survey is conducted face-to-face to avoid the misunderstanding of the questions.

4.2.2.2 Site Observation

As a favored method for portraying neighborhood environment, neighborhood observation is applied to collect information on factors influencing health. Site visits were conducted on weekdays and weekends, from 3:30 pm to 7:30 pm from March 16 to March 23, 2015. The physical conditions of the residential environment of the three cases and residential behaviors are observed. The site observation focused on the land use pattern of neighborhood, travel mode of residents, urban design characteristics (landscape and facilities), usage of public transportation, usage of green and open space and service facilities (especially the sport facilities), daily recreational activities, social interactions between neighbors, and community activities.

4.2.2.3 Semi-Structured Interview

Semi-structured interview was conducted to collect qualitative data for explaining influential mechanisms. Two rounds of interview were held from November 15 to December 5, 2015 and from June 2 to June 9, 2016. Interview questions were based on the results of quantitative analyses, focusing on the factors influencing health the residents. Six groups of stakeholders were interviewed, including residents from the three ceases, developers in Guangzhou, Guangzhou government officials, planners with experience in housing planning and design, housing managers and the heads of residents' committees of the three SCHEs. Specifically, the residents were interviewed with the perceived effects of suburban living on their health; the officials were interviewed with the regulations and interventions on the housing development; the developers were interviewed with decision-making of the housing development, focusing on the strategies for suburban housing development and the influential factors of the physical residential environment; the planners were interviewed in terms of the professional techniques that are beneficial for health-promoting environment; the housing managers and residents' committee were interviewed in terms of the affairs in housing management. Finally, there were 40 interviewees, comprised of 23 residents (targeted: 28 residents), four developers (targeted: 3 developers), two officials (targeted: 4 officials), six planners (targeted: 6 planners), three housing managers (targeted: 3 housing managers) and two staffs in residents' committees (targeted: 3 staffs). Although several targeted numbers were not achieved, the representativeness of the interviews was basically satisfied since different groups of stakeholders holding distinctive opinions from others were interviewed, which is more important than the various opinions within the groups. The detailed interview questions and interviewee list are in Appendix 3.

4.2.2.4 Desktop Research

Desktop research is done to collect secondary data to understand the contextual factors of SCHEs development in Panyu. Regulatory and site plans and macro context of SCHEs development (urban policies and economic and socio-cultural contexts) were collected. The materials consist of:

1. Regulatory plan of Panyu District.
2. Site plans of the selected SCHEs.
3. The latest master plan and conceptual development plan of Panyu District.
4. The latest master plan and conceptual development plan of Guangzhou.
5. Related ordinances, regulations, guidelines for residential planning and design, such as Town Planning Ordinance of Guangzhou, Code for planning and design on urban residential areas (GB50180-93, 2002 version), design code for residential building (GB50096-2011).
6. Statistical yearbook of Panyu District (1998–2013).
7. The latest census data in Panyu District.
8. Housing policy and planning policy.
9. Publications, reports, and studies carried out by governmental departments.

4.2.3 Quantitative Analysis Methods

First of all, the data collected in the questionnaire survey was cleaned and analyzed. Unfortunately, there were high numbers of missing value in terms of depression scales (20 items, 46.9% of total score was missing) and income (17.8%). Based on the face-to-face investigation, it could be inferred that the missing data could have resulted from carelessness or impatience of respondents and the willing of protecting personal privacy, which is the limitation in the questionnaire survey. Listwise deletion was adopted since there is no good way to analyze data set missing at non-random. For other variables with up to 5% of missing data, listwise deletion has been used. Although listwise deletion may yield biased parameter estimates, the bias is often minimal as long as those variables can reasonably be involved in the model as covariates (Graham [9], as cited in Graham and Donaldson 1993).

For the quantitative data in this study, statistical analyses are performed with SPSS statistical software. In terms of the evaluation of health performance, four variables have been developed according to the discussion in Sect. 4.1.1; their measurement and data type are shown in Table 4.8. Descriptive analysis, chi square and one-way ANOVA were adopted in the first step of analysis. Next, multiple regression models are used to examine the associations between the health performance and the potential factors influencing health, including the residential environment and health-related residential behaviors. To operate, three groups of independent variables for the health performance are investigated: (1) 17 variables of perceived residential environment factors stemming from the four domains of transport environment, leisure environment, service environment, and design features (Table 4.9); (2) 18 variables of health-related residential behaviors stemming from three aspects of physical activity, neighbors interactions, and residents' participation (Table 4.10); and (3) socio-demographic characteristics (age, gender, income, education, employment, homeownership, residence length, and family structure). With the hypothesis that all of these factors could affect health outcomes, the associations between the independent variables and the health performance are analyzed, through which the factors influencing health could be identified.

However, a methodological caveat has to be raised in exploration of determinants of mental health. It is tricky to take the subjective measurements for the residential environment and health-related residential behaviors, because the assessment may suffer from the reporting or same-source bias, confounding the identification of the cause. The reporting bias may arise when people diagnosed with depression report a worse external environment because of their depression [16]. It is very likely that depressed respondents feel more negative assessment about their environment, even though its quality is objectively very good. So it is difficult to differentiate the bias from the real reasons for the influential mechanism. In this context, for the factors affecting depression, only the objective ones that are not likely to be the consequence of the depression are used in the regression analyses. For the factors that influence RMD, since the symptoms of low score could not be diagnosed with mental illness, it is assumed that the subjective assessment on residential environment and health-related residential behaviors are less likely to be affected by RMD. In order to exclude

Table 4.8 Dependent variables and the measurement

Domain	Variables	Measurement	Type of data
Physical health	Self-report general physical health condition	How do you feel your general physical health? (Excellent, very good, good, fair, poor)	Ordinal
	Obesity	Body mass index	Binary
Mental health	Depression disorder	CES-D scale	Binary
	Risks of mental disorders (RMD)	A three-item scale How much of the time have you felt anxious during the past 4 weeks? How much of the time have you felt stressful during the past 4 weeks? How much of the time have the mood of anxiety and stress affected your daily residential living?	Interval
		(Average value: summation of the above three variables divided by 3)	

Table 4.9 Variables of perceived residential environment and the measurement

Domain	Variables	Measurement	Type of data
Transportation environment (Walking environment)	Perception of accessibility of destinations by walking	How do you perceive the convenience of walking to (1) subway station, (2) bus stop, (3) retail and market, (4) leisure facilities, (5) kindergarten and primary school, (6) clinic or pharmacies, (7) community services? *(Average value: summation of the seven sores divided by 7, ranging from 1–5)*	Interval
	Perception of the quality of walking surface	The walking surface (pavement, street lamp, trees) is suitable for walking *(Strongly disagree, disagree, not sure, agree, strongly agree)*	Ordinal
Leisure environment	Place preference for leisure activity	Where do you usually do above leisure activities? *(Within neighborhood, Outsider neighborhood, Others)*	Nominal
	Perception of the quantity of open space	There are enough outdoor open space (square, walking path) to do exercises *(Strongly disagree, disagree, not sure, agree, strongly agree)*	Ordinal
	Perception of the quantity of fitness facilities	There are not enough fitness facilities (playing court, swimming pool, fitness equipment) within the neighborhood *(Strongly disagree, disagree, not sure, agree, strongly agree)*	Ordinal
	Satisfaction with the quality of space and facilities	I am satisfied with the quality of the open space and fitness facilities for physical exercises *(Strongly disagree, disagree, not sure, agree, strongly agree)*	Ordinal
Service environment	Places chosen for health issues	For general sickness, where will you go for a checkup? (*Community clinics, Hospital in Panyu district, Hospital in urban core area, Others*)	Nominal
	Satisfaction with education facilities	The schools within this neighborhood could not satisfy my demand. *(Strongly disagree, disagree, not sure, agree, strongly agree)*	Ordinal
	Satisfaction with health care facilities	There are not enough health facilities, which could not satisfy my demand *(Strongly disagree, disagree, not sure, agree, strongly agree)*	Ordinal
	Satisfaction with daily commercial facilities	There are not enough commercial facilities, which could not satisfy my demand *(Strongly disagree, disagree, not sure, agree, strongly agree)*	Ordinal

(continued)

Table 4.9 (continued)

Domain	Variables	Measurement	Type of data
Intangible environment	Perceived safety	The neighborhood is safe from crime *(Strongly disagree, disagree, not sure, agree, strongly agree)*	Ordinal
	Perceived density	I feel the population density is all right, neither too crowd or too deserted *(Strongly disagree, disagree, not sure, agree, strongly agree)*	Ordinal
	Perception of green space	The green space (including trees and grass) is not enough in this neighborhood *(Strongly disagree, disagree, not sure, agree, strongly agree)*	Ordinal
	Perception of ambient environment	The ambient environment makes me comfortable (good air quality, good acoustic environment, good hygiene condition *(Strongly disagree, disagree, not sure, agree, strongly agree)*	Ordinal
	Perception of public realm	There are enough social space within the neighborhood for neighbors' interactions and community activities *(Strongly disagree, disagree, not sure, agree, strongly agree)*	Ordinal
	Sense of aesthetic	The appearance of the landscape and architecture are characteristic and looks good to me *(Strongly disagree, disagree, not sure, agree, strongly agree)*	Ordinal
	Sense of belonging	The design of this neighborhood makes me have a feeling of attachment and belonging *(Strongly disagree, disagree, not sure, agree, strongly agree)*	Ordinal

the effects of depression, the answers given depressed respondents are not considered in the analyses of influential factors of RMD.

Furthermore, qualitative data collected from interviews and observations are used to draw conclusions about the possible influencing mechanisms and to reveal the phenomena that are invisible in statistical analysis. Content analysis is adopted in the interpretation of the findings of the statistical analysis. In particular, interview findings are linked with the associations between different groups of factors, integrating different data to validate the findings of this study.

Table 4.10 Variables of health-related residential behaviors and the measurement

Domain	Variables	Measurement	Type of data
Travel activity	Travel mode for commuting	Which means of transport do you usually use when you go to work, daily shopping and for recreation (rest and exercises)? (*Active travel mode, Public transportation, Private car, Other or multiple choice*)	Ordinal
	Travel time	How much time do you usually spend traveling on an ordinary day (sum of travel to work, shopping and recreation)? (*< 1.5 h, 1.5–2 h, 2–3 h, > 3 h*)	Ordinal
	Time spend on walking	How much time do you usually spend walking on an ordinary day (sum of travel to work, shopping and recreation)? (*< 30 min, 30–59 min, 60–89 min, 1.5–2 h, > 2 h*)	Ordinal
	Willingness for walking	Compared to other means, I prefer to walk to destinations around the neighborhoods (*Strongly disagree, disagree, not sure, agree, strongly agree*)	Ordinal
Leisure activity	Variety of leisure activities	Numbers of leisure activities you participated in during the past six months (*No leisure activity, one type, two types*)	Ordinal
	Preference for leisure activity	I love doing leisure activities such as physical exercises (*Strongly disagree, disagree, not sure, agree, strongly agree*)	Ordinal
	Regularity of doing leisure activities	I did physical exercise regularly over past six months (*Strongly disagree, disagree, not sure, agree, strongly agree*)	Ordinal
Informal neighbors interaction	Variety of informal interaction	How many types of informal interactions have you usually engaged in over the last six months? (*One type and less, Two types, Three types and more*)	Ordinal
	Frequency of informal interaction	How often do you interact with your neighbors informally over past six months? (*Everyday, More than 3 times per week, Once or twice per week, Once per month, Never*)	Ordinal
	Willingness to join informal neighbors' interactions	In my spare time, I'd rather to stay at home than to interact with my neighbors (*Strongly disagree, disagree, not sure, agree, strongly agree*)	Ordinal

(continued)

Table 4.10 (continued)

Domain	Variables	Measurement	Type of data
Community activities	Variety of community activities	How many types of community activities have you usually engaged in during last six months? *(None, One type, Two types and more)*	Ordinal
	Frequency of community activities	How often did you attend community activities in the past six months? *(Never, Once per month or less, More than once per month)*	Ordinal
	Willingness to attend community activities	I have no interests in community activities *(Strongly disagree, disagree, not sure, agree, strongly agree)*	Ordinal
	Perceived neighborhood cohesion	Neighborhood cohesion scale (Residents are willing to help their neighbors; Residents in this neighborhood can be trusted; This is a close-knit and strong social inclusive neighborhood; Residents in this neighborhood generally don't get along with each other; The relationships between residents of this neighborhood and surrounding villagers are good) *(Average value: summation of the five scores divided by 5, ranging from 1 to 5)*	Interval
Participation in housing operation	Variety of participation	How many types you usually participate in the housing operation affairs? *(None, None type, Two types and more)*	Ordinal
	Frequency of participation	How often did you participate the housing operation affairs over past six months? *(Never, Less than once per month, Once per month, More than once per month)*	Ordinal
	Concerns for neighborhood development	I don't care the housing operation and related affairs at all *(Strongly disagree, disagree, not sure, agree, strongly agree)*	Ordinal
	Satisfaction with housing operation	I am not satisfied with the housing operation *(Strongly disagree, disagree, not sure, agree, strongly agree)*	Ordinal

4.3 Conclusion

This chapter proposes an analytical framework to answer the three research questions. The health performance of residents in SCHEs is evaluated from the dimensions physical and mental health, based on which the variation across the three SCHEs are analyzed. The associations between the health of residents and residential environment, health-related residential behavior are examined to ascertain the factors influencing health. Factors from the residential environment are examined in both objective and subjective ways, representing physical residential environment and perceived residential environment, respectively. The set of factors influencing health affecting the health of residents could be explained by examining the residents-environment transactions from the human ecological perspective, and by investigating the institutional factors in suburban housing development.

This chapter also describes the research methods. Three housing estates in Panyu District in Guangzhou are selected as the study cases. While both qualitative and quantitative methodologies are adopted, according to which the data collection method is applied: questionnaire survey, site observation, semi-structured interview and desktop research. Consistent with the analytical steps, statistical analyses methods are referred to such as Chi-square, one-way ANOVA, and multiple regression. Finally, the dependent variables and independent variables for the regression analysis are developed and listed in two tables, which form a significant foundation for the second step of analysis.

References

1. Bhatti M (1994) Environmental futures and the housing question. In: Bhatti M, Brooke J, Gibson M (eds) Housing and the environment: a new agenda. Chartered Institute of Housing, Coventry, pp. 14–33
2. Bonaiuto M, Fornara F, Bonnes M (2003) Indexes of perceived residential environment quality and neighbourhood attachment in urban environments: a confirmation study on the city of Rome. Landsc Urban Plan 65(1):41–52. https://doi.org/10.1016/S0169-2046(02)00236-0
3. Bonaiuto M, Fornara F, Bonnes M (2006) Perceived residential environment quality in middle- and low-extension Italian cities. Eur Rev Appl Psychol 56(1):23–34. https://doi.org/10.1016/j.erap.2005.02.011
4. Carp FM, Zawadski RT, Shokrkon H (1976) Dimensions of urban environmental quality. Environ Behav 8(2):239–264. https://doi.org/10.1177/001391657682004
5. Chiu RLH (2000) Environmental sustainability of Hong Kong's housing system and the housing process model. Int Plan Stud 5(1):45–64. https://doi.org/10.1080/135634700111819
6. Eicher C, Kawachi I (2011) Social capital and community design. In: Dannenberg AL, Frumkin H, Jackson RJ (eds) Making healthy places: designing and building for health, well-being, and sustainability. Island Press/Center for Resource Economics, pp. 117–128. https://doi.org/10.5822/978-1-61091-036-1_8
7. Felce D, Perry J (1995) Quality of life: its definition and measurement. Res Dev Disabil 16(1):51–74. https://doi.org/10.1016/0891-4222(94)00028-8
8. Frank L, Engelke P, Schmid T (2003) Health and community design: the impact of the built environment on physical activity. Island Press

9. Graham JW (2009) Missing data analysis: making it work in the real world. Annu Rev Psychol 60(1):549–576. https://doi.org/10.1146/annurev.psych.58.110405.085530
10. He JJ, Jia B (2007) Sustainable residential landscapes: a case study in Guangzhou. China. Landscape Research 32(2):241–254. https://doi.org/10.1080/01426390701231747
11. Hu W (2007) Residential area planning principle and design method. China Architecture & Building Press
12. Jackson RJ, Kochtitzky C (2003) Creating a healthy environment: the impact of the built environment on public health. Sprawl Watch Clearinghouse Monograph Series, 19.
13. Li L (2007) A home of one's own: ways of life in privately governed commodity housing estates in urban China. ProQuest
14. Li P (2008) The reform and opening up of Panyu. Nanfang Daily Press
15. Lu Y, Hu P, Treiman DJ (2012) Migration and depressive symptoms in migrant sending areas: findings from the survey of internal migration and health in China. Int J Public Health 57(4):691–698. https://doi.org/10.1007/s00038-011-0314-0
16. Mair C, Roux AVD, Galea S (2008) Are neighbourhood characteristics associated with depressive symptoms? a review of evidence. J Epidemiol Community Health 62(11):940–946. https://doi.org/10.1136/jech.2007.066605
17. Marans RW (2003) Understanding environmental quality through quality of life studies: the 2001 DAS and its use of subjective and objective indicators. Landsc Urban Plan 65(1):73–83. https://doi.org/10.1016/S0169-2046(02)00239-6
18. Moser G (2009) Quality of life and sustainability: toward person–environment congruity. J Environ Psychol 29(3):351–357. https://doi.org/10.1016/j.jenvp.2009.02.002
19. Northridge ME, Sclar ED, Biswas P (2003) Sorting out the connections between the built environment and health: a conceptual framework for navigating pathways and planning healthy cities. J Urban Health 80(4):556–568. https://doi.org/10.1093/jurban/jtg064
20. Pacione M (2003) Urban environmental quality and human wellbeing—a social geographical perspective. Landsc Urban Plan 65(1):19–30. https://doi.org/10.1016/S0169-2046(02)00234-7
21. Paulhus DL, Vazire S (2007) The self-report method. Handbook of research methods in personality psychology 1(2007):224–239
22. Radloff LS (1977) The CES-D scale: a self-report depression scale for research in the general population. Appl Psychol Meas 1(3):385–401. https://doi.org/10.1177/014662167700100306
23. Roberts RE, Vernon SW (1983) The center for epidemiological studies depression scale: its use in a community sample. Am J Psychiatry 140:41–46. https://doi.org/10.1176/ajp.140.1.41
24. Shen Y-C, Zhang M-Y, Huang Y-Q, He Y-L, Liu Z-R, Cheng H, Tsang A, Lee S, Kessler RC (2006) Twelve-month prevalence, severity, and unmet need for treatment of mental disorders in metropolitan China. Psychol Med 36(2):257–267. https://doi.org/10.1017/S0033291705006367
25. Shin S-C, Kim M-K, Yun K-S, Kim J-H, Lee M-S, Moon S-J, Lee M-J, Lee H-Y, Yoo K-J (1991) The center for epidemiologic studies-depression scale: its use in Korea. J Korean Neuropsych Assoc, 752–767
26. Singleton R Jr, Straits BC, Straits MM (1993) Approaches to social research. Oxford University Press
27. Stokols D (1992) Establishing and maintaining healthy environments: toward a social ecology of health promotion. Am Psychol 47:6–22. https://doi.org/10.1037/0003-066X.47.1.6
28. Stokols D (1996) Translating social ecological theory into guidelines for community health promotion. Am J Health Promot 10(4):282–298. https://doi.org/10.4278/0890-1171-10.4.282
29. Stokols D, Altman I (1987) Environmental psychology. Percept Cognitive Image City 12:23–90
30. Taylor SE, Repetti RL, Seeman T (1997) Health Psychology: what is an unhealthy environment and how does it get under the skin? Annu Rev Psychol 48(1):411–447. https://doi.org/10.1146/annurev.psych.48.1.411
31. Turcu C (2013) Re-thinking sustainability indicators: local perspectives of urban sustainability. J Environ Planning Manage 56(5):695–719. https://doi.org/10.1080/09640568.2012.698984
32. van Kamp I, Leidelmeijer K, Marsman G, de Hollander A (2003) Urban environmental quality and human well-being: towards a conceptual framework and demarcation of concepts; a

literature study. Landsc Urban Plan 65(1):5–18. https://doi.org/10.1016/S0169-2046(02)00232-3
33. Verheij RA (1996) Explaining urban-rural variations in health: a review of interactions between individual and environment. Soc Sci Med 42(6):923–935. https://doi.org/10.1016/0277-9536(95)00190-5
34. Walmsley DJ, Lewis GJ (2014) People and environment: behavioural approaches in human geography. Routledge
35. Wu F, Shen J (2015) Suburban development and governance in China. In: Suburban governance: a global view. University of Toronto Press

Chapter 5
Health Performance of Residents in SCHEs

The expeditious urbanization in China is accompanied with urban spatial expansion and lifestyle transition, contributing to the health transition from infectious diseases to chronic diseases. The twenty years has seen an obvious and speedy increase in the prevalence of obesity of Chinese population, leading to chronic and cardiovascular disease. In particular, the middle-income population is getting fat fast with the changes of diet and lifestyle, especially the physical activity. For mental health, the prevalence of mental disorders of Chinese people has been underestimated, including both clinical disorder and everyday psychopathology such as anxiety and stress. In addition, these health implications of suburbanization have been less discussed in Chinese context. Thus, this chapter presents the findings of the analysis on the general health of the residents in SCHEs.

This chapter answers the first research question by evaluating the physical and mental health of residents in the three SCHEs. Two indicators for physical health are assessed: body mass index (BMI) and self-reported general physical health condition. Two indicators of mental health are measured: depression and risks of mental disorders (RMD). In addition to the descriptive analyses of the health performance, comparisons across the three SCHEs are made to determine the effects of the living environment of three neighborhood types on residents' health. Finally, the interrelations between physical health and mental health are then examined, based on which conclusion is made.

T. Zhang, *Towards Healthy Settlements*, Urban Sustainability,
https://doi.org/10.1007/978-981-97-1207-6_5

5.1 Physical Health Performance of Residents in SCHEs

5.1.1 Prevalence of Obesity

As explained in chapter 4, BMI is used to diagnose obesity. For data collection, respondents were initially asked to write down their height and weight, based on which the BMI was calculated. The reliability of self-reported data collection is questionable since people are not always honest about their height and weight. For this reason, 10 handheld scales were carried into the survey sites, and the respondents were requested to record their height and weight in privacy. The scales then calculated the BMI. Although this was still a self-reported way of collecting data, it was assumed that it would be more accurate since respondents gave the data in a relative private way. This method generated a calculated BMI by the interviewer and a measured BMI by the scale. To test the hypothesis that the calculated BMI ($M = 22.07$, $SD = 2.81$) and measured BMI means ($M = 22.11$, $SD = 2.98$) were equal, a paired sample t-test was performed. The correlation between the two variables was estimated at $r = 0.872$, $p < 0.001$, suggesting that the method of paired samples t-test was appropriate for above purpose. The null hypothesis of equal BMI means was not rejected, $t\,(378) = 0.602$, $p = 0.548$. This indicates that the calculated BMI was equal to the measured BMI. So the following analysis uses the calculated BMI since there was less missing data than the measured BMI.

In terms of the benchmark for adjusting obesity, studies point out that the WHO guideline (obesity: BMI $> = 30.0$ kgm^{-2}; overweight: BMI $> = 25.0$ kgm^{-2}) may understate the true situation (Wu, 2006). In this study, "Chinese adult overweight and obesity prevention and control guidelines" are used and the criterion for diagnosing the obesity is a BMI of at least 25.0 kgm^{-2}. As a result, the BMI scores of respondents suggested that 11.1% were underweight (BMI $<$ 18.5 kgm^{-2}), 56.0% were within the normal range (BMI ranges from 18.5 kgm-2 to 22.9 kgm^{-2}), 17.6% were overweight (BMI $> = 23.0$ kgm^{-2}), and 15.3% respondents were obese (including moderate ones and severe ones) (Fig. 5.1). Compared to previous studies, the percentage of obesity was similar to that of the general population of China (14.7%) in 2002, based on health survey data (Wu, 2006). In this sense, it could be inferred that the measurement of obesity here is reliable.

5.1.2 Self-Reported General Physical Health

Self-reported general physical health was measured through a five-scale item with the question "How would you describe your general physical health during the past four weeks?" Of the respondents 71.7% reported that they were in good physical health, 25.4% said that their physical health was fair, and only 2.9% admitted they were in poor physical health (Fig. 5.2).

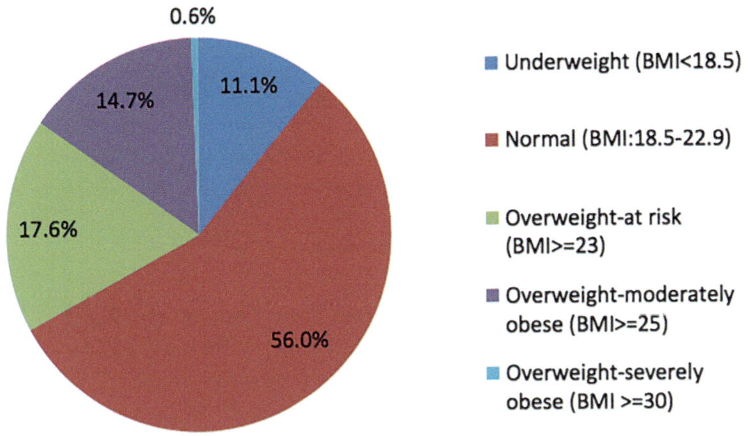

Fig. 5.1 Body weight status proportion based on BMI (N = 523)

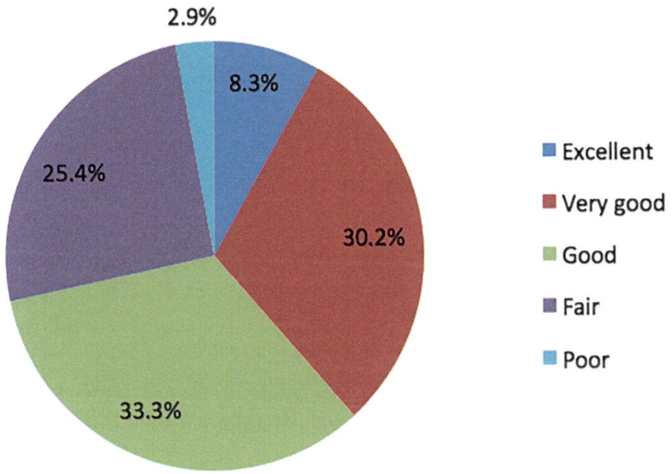

Fig. 5.2 Self-reported general physical health (N = 520)

The influences of general physical health on daily life were investigated through the question: "During the past four weeks, have you had any of the following problems with your work or other regular daily activities as a result of your physical health?" As a result, 21.9% (N = 527) residents reported that they accomplished less than they would like, and 13.3% (N = 527) reported that they were restricted in performing the kind of work or other activities as they intended.

According to the interviews with different populations, differences were found in the self-reported assessment of physical health between the healthy elderly and healthy young people. The elderly has much more time for exercises and less stress than young people, leading to better perception of physical health. Two residents remarked as follows.[1]

We elderly are very healthy actually. Other than taking care of the grandchildren, we spend much time on physical exercise and recreational activities such as singing, dancing, and playing cards. We have little to worry and we could do whatever we like to do. However, the young people suffer more stress than us and they have little time to do exercise. (AG_MO6).

5.1.2.1 I Think I Am Unhealthy Although I Am Only in My 30 s. I Suffer Much from the Stress in the Daily Life. (AG_MY7)

To summarize, around 28.3% of respondents did not report good general physical health, including those felt fair (25.4%) and those reported poor physical health (2.9%). While 21.9% reported that they accomplished less due to the physical health issues, indicating that around one-fifth residents could be affected by the physical health problems in their daily life. Thus, it is inferred that the respondents were less likely to be severely ill during the survey period. For healthy residents, the elderly generally reported better physical health than the younger people.

5.1.3 Physical Health Across the SCHEs

A chi-square test was conducted to examine the independence of the categorical BMI across the three SCHEs (Table 5.1). Although the statistics showed that the proportion of the overweight in AG was the lowest among the three neighborhoods numerically, the results of chi-square test suggested no significant difference in proportion across the three neighborhoods, X^2 (8, N = 523) = 4.876, $p = 0.771$. Thus, proportionally, no significant difference of body weight was found across the three SCHEs in this study.

Similarly, the proportion difference of self-reported general physical health was analyzed through a chi-square test (Table 5.1). Although the proportion of respondents reporting good physical health in AG was the highest numerically, the null hypothesis that the self-reported general physical health of residents in the three SCHEs were equal was not rejected, X^2 (8, N = 520) = 7.049, $p = 0.531$. Statistically, there was no significant difference in proportion in general physical health across the three SCHEs.

Since no significant difference across the three SCHEs was found in body weight and self-reported general physical health, this study could not prove the proposition

[1] The socio-economic characteristics of the residents in the three neighborhoods were similar, and it is found that residents from different neighborhoods shared similar opinions in most interview questions during the interview. Thus, in terms of interpreting the opinions, interviewees were treated as individual, who was not influenced by where they live.

Table 5.1 Physical health by SCHEs

	LNT (N = 174)	AG (N = 208)	GV (N = 145)
BMI (categorized) [a]			
Underweight	9.8 (17)	14.6 (30)	7.7 (11)
Normal	56.9 (99)	54.9 (113)	56.6 (81)
Overweight-at risk	17.8 (31)	16.0 (33)	19.6 (28)
Overweight-moderately obese	14.9 (26)	14.1 (29)	15.4 (22)
Overweight-severely obese	0.6 (1)	0.5 (1)	0.7 (1)
Total	174	206	143
Self-reported general physical health [b]			
Excellent	8.7 (15)	8.3 (17)	8.3 (11)
Very good	31.8 (55)	31.6 (65)	26.2 (37)
Good	30.1 (52)	35.9 (74)	33.3 (47)
Fair	26.0 (45)	20.9 (43)	31.2 (44)
Poor	3.5 (6)	3.4 (7)	1.4 (2)
Total	173	206	141

Notes: Percent (number of cases)
[a] X^2 (8, N = 523) = 4.876, $p = 0.771$
[b] X^2 (8, N = 520) = 7.049 $p = 0.531$

that neighborhood type affects physical health. Conversely, it is proposed that the physical health of residents may not be influenced by the differences of physical environment across the three neighborhood types. This will be further explained in chapter 6 when explore the factors affecting physical health. In addition, to find out the reasons for the similarity of physical health across the three SCHEs, the following passage interprets residents' opinions on the elements that influence their perception of physical health.

The opinions of the residents were collected (Table 5.2). The most frequently mentioned components affecting physical health were mood, diet, quality of the living environment, and habits or physical activity. Residents of different SCHEs mentioned similar determinants of physical health; for the quality of the living environment, residents always referred to the quality of the natural and ambient environment, such as the air, water, and even weather (Table 5.2). Thus, since the three types of SCHEs enjoy a similar natural environment, the physical residential environment of SCHEs at the micro level was unlikely to have obvious effects on physical health. Therefore, this study proposes that the physical residential environment of the three neighborhood types has little influence on the physical health of its residents.

Table 5.2 Statements on crucial factors influencing physical health

	Mood	Diet	Living environmental quality	Living habit (physical activity)
LNT				
Environmental quality is the most important. Since there are many factories in Panyu district, the water and air are not very good. Then, the diet and daily routine are also could impact the physical health. (LNT_FY1)		✔	✔	✔
Diet and stress. Since I have been living in this area since I was very young, I didn't feel any effects of the environment quality on my health. (LNT_FM2)	✔	✔		
I think the most important three factors are environment, mood, and diet. (LNT_FM5)	✔	✔	✔	
Regular daily routine is important for health, and then the environment. (LNT_MM6)			✔	✔
Mood is the most important one, it is usually affected by work, social relations and weather. (LNT_FY7)	✔			
The effects of emotional status are crucial. Many people living in big cities suffer from stress, which is a potential risk for health. And living environment could influence the emotion as well, for example, the air pollution could make people feel bad. For another, healthy diet and regular daily routine are important for health. You should do the right thing at the right time. (LNT_FM8)	✔		✔	✔

(continued)

Table 5.2 (continued)

	Mood	Diet	Living environmental quality	Living habit (physical activity)
Stress, mood, and living environment. If I come back home, where is full of singing birds and of sweet-scented flowers, I will feel very good. And the air quality and the service facilities could also influence my health. If there is a garden or some green and open space, I think I will do more physical exercise. (LNT_MY9)	✔		✔	
AG				
Daily routine, stress, diet, physical exercise, and living environment. (AG_FM1)	✔	✔	✔	✔
I think there are two important factors for me: good psychology and physical exercises. Generally, diet and physical exercise are the most influential factors for health, but I eat moderately and wisely. In addition, keeping early hours is also important for keeping health. (AG_MY3)	✔	✔		✔
Mood, air quality, physical exercise, and personal trait. (AG_FY4)	✔		✔	✔
GV				
Should keep in good mood and have a healthy diet. (GV_FO1)	✔	✔		
Good mood and air quality are the main factors influencing my physical health. (GV_FM2)	✔		✔	

(continued)

Table 5.2 (continued)

	Mood	Diet	Living environmental quality	Living habit (physical activity)
Go to bed earlier and watch less TV could help to keep healthy. And I like to walk around and to dance with neighbors, which could make me feel better and have a better appetite. Good mood is important for physical health. (GV_FO4)	✔			✔
I am not sure about the influential factors, but I like to go for a walk after dinner. (GV_FO6)				✔

5.2 Mental Health Performance of Residents in SCHEs

5.2.1 Depressive Disorder

Depressive disorder was evaluated based on the CES-D scale. However, within the sample (N = 527), the response rate was 54.8%, with 289 valid cases. Considering that this is personal sensitive information, the response rate is acceptable. Descriptive statistics showed that the depression score varied from 0 to 60, with higher score implying more serious depressive symptoms.

With regards to the designation of a cutoff point for depressive symptom confirmation, Western countries and Asia have adopted 16 as an optimal CES-D cutoff point [1], as cited in Shima et al.1985,[9, 10]. Adopting the recommended cut-off score of 16, 12.1% respondents (N = 35) were diagnosed as suffering from depression, while 87.9% respondents were not. The prevalence of depression in this study was similar with that in previous studies conducted in China. For example, [7] found that there were around a quarter of the sample (migration) exhibiting depressive symptomatology with the same depression measurement. For another, with the CES-D measurement, it is found that the prevalence of depression in general population from Beijing was 14.9% while that in general population from Shanghai was 4.1% [8]. Thus, studies conducted in different geographic locations and focusing on different groups of populations make some differences in the prevalence of depression, bringing broads insights in understanding of depression.

Across the three SCHEs, the percentage of people diagnosed as depressed was highest in AG (23.7%); while only 5.3% of residents in LNT and 2.6% in GV were diagnosed as depressed (Fig. 5.3). To figure out the independence of the prevalence of depression across the three SCHEs, chi-square test was performed. Results suggested that the difference in proportion across the three SCHEs was statistically significant,

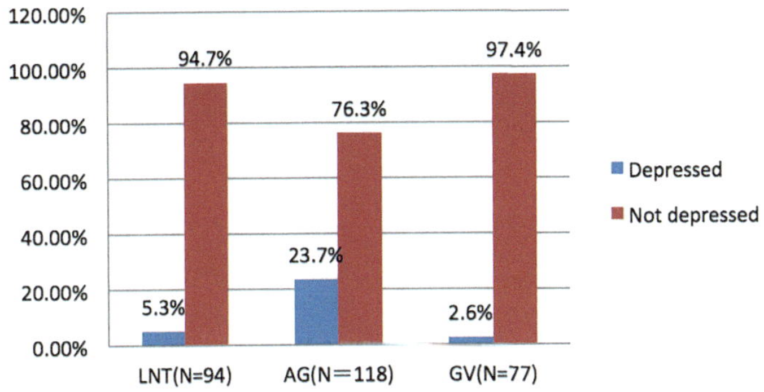

Fig. 5.3 The percentage of depression by SCHEs (N = 289)

X^2 (2, N = 289) = 25.584, $p < 0.001$. Thus, the prevalence of depression in AG was much higher than that in LNT and GV. Hence, unlike physical health, there is significant difference in mental health among the three selected SCHEs.

5.2.2 *Risks of Mental Disorders*

A three-item five-point scale was developed to measure the risk of mental disorders, focusing on anxiety, stress, and their effects on daily living. The questions of "How much time have you felt anxious/stressed/affected by anxiety and stress in daily residential living during the past four weeks?" were asked, and respondents could choose one of the five answers [15] (Table 5.3). In terms of the internal consistency reliability of the scale, the Cronbach's alpha[2] is 0.713 (N = 527), suggesting that the three items could reliably measure the risks of mental disorders. Then a comprehensive variable, risks of mental disorders (RMD), was generated by aggregating the three items with the same weight, calculated by the sum of the three items divided by 3. The RMD score ranges from 1 to 5, with the mean of 3.98 (Table 5.3).

The pattern of the risks is as follows. For anxiety ($M = 3.93$, $SD = 0.95$), 30.1% respondents felt none, and 62.0% respondents admitted feeling a little or some, and 7.9% respondents reported feeling more or much more. For stress ($M = 3.73$, $SD = 1.16$), 30.3% respondents felt none, 53.7% felt some or a little, and 16.0% respondents reported that they felt more or much more. For the effects of the anxiety and stress on residential living in the past four weeks ($M = 4.29$, $SD = 0.93$), 52.0% respondents reported that they had felt none, 41.0% felt some or a few effects, and 7.0% felt more or much more effects (Fig. 5.4).

For the judging criteria of RMD, literature suggest that an average score of less than 3 could be taken as a benchmark for high risks of mental disorders [13–15], according to which this study suggested that risks of mental disorders were relatively

[2] If Cronbach's alpha is higher than 0.7, the reliability of the scale is acceptable.

Table 5.3 Descriptive analysis of mental health

Item	Scale	M	SD
Level of anxiety [a]	None (5) Little (4) Some (3) More (2) Much more (1)	3.93	0.94
Level of stress [a]	None (5) Little (4) Some (3) More (2) Much more (1)	3.72	1.18
Effect of anxiety and stress on daily residential living [a]	Much more (1), More (2), Some (3), Little (4), None (5)	4.28	0.94
Risk of mental disorders [a]	Mean of the three items	3.98	0.82

Note: a. Ranging from 1–5; a higher value means less risk on mental disorders

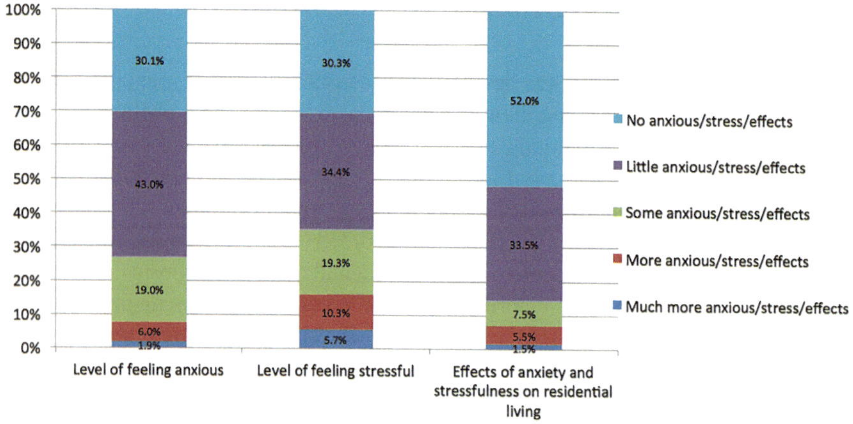

Fig. 5.4 Assessment of three risks of mental disorders (N = 527)

low (3.98 > 3). When taking the status of "more anxiety, more stress, more effects" as benchmarks for high risks of mental disorders (less than 3), it was found that 7.9% of respondents had severe anxiety, 16.0% had severe stress, and 7.0% reported severe negative effects on daily residential living.

Across the three cases, numerically, GV had the highest mean score of RMD ($M = 4.20$, $SD = 0.77$) among the three SCHEs, and the scores of AG and LNT are very close (Table 5.4). To examine whether the mean differences across the three neighborhood types are significant, a one-way ANOVA test was performed. Prior to the analysis, the assumption of normality was evaluated and determined to be

Table 5.4 One-way ANOVA of RMD

	N	M	SD	Skew	Kurtosis
LNT	174	3.84	0.78	-0.705	1.017
AG	208	3.93	0.85	-0.766	0.268
GV	145	4.20	0.77	-0.884	0.182

Note: range for mean of RMD is from 1 to 5

Table 5.5 Results associated with Tukey's HSD multiple comparisons

Comparison	*p*	*Cohen's d*
LNT-AG	0.539	-0.11
LNT-GV	0.000	-0.46
GV-AG	0.007	-0.33

Note: According to Cohen's guideline, d = 0.2 is a small effect, d = 0.5 is a moderate effect, d = 0.8 or > is a large effect

satisfied, and the assumption of homogeneity of variances was tested and satisfied, Levene's F test, F (2, 524) = 1.430, $p = 0.240$. Thus, the independent one-way ANOVA yielded a statistically significant effect, F (2, 524) = 8.084, $p < 0.001$ (Table 5.4). The null hypothesis of no difference between the means was rejected, indicating that the risks of mental disorders varied significantly across the three SCHEs.

Furthermore, to evaluate the differences among the three means, Tukey's HSD multiple comparisons were performed. According to Table 5.5, comparison between LNT and GV and comparison between AG and GV were both statistically significant ($p < 0.05$), indicating that GV performed better than both LNT and AG significantly. According to Cohen's guidelines, the statistically significant difference between the means were associated with small to moderate effect sizes [11].

To summarize, many more residents in AG than in LNT and GV were diagnosed with depression. In terms of the RMD, there was no difference between living in AG and LNT, but the respondents in GV had much lower risks of mental disorders than those living in LNT and AG. Thus, residents in GV may be least vulnerable to mental disorders, which will be explored further in Chapter 6.

5.3 Interactions Between Physical and Mental Health

As literature suggests that physical health and mental health are closely interwoven and deeply interdependent [12], the interactions between the two dimensions deserve attention. Different symptoms co-vary together, meaning that people reporting physical symptoms (headaches, stomach-aches, nausea) are very likely to report mood disturbances such as feeling sad, hopeless and lonely [2]. Thus, physical health could be both the reasons and results of mental health, and vice versa.

Many interviewees mentioned the obvious effects of moods on physical health. In order to examine the proposed associations between physical and mental health, bivariate correlation analysis was conducted between variables of self-reported general physical health and the two mental health variables. Spearman's rho showed that RMD and self-reported general physical health correlated significantly, r_s (527) $= 0.095$, $p = 0.030$, suggesting that better mental health was associated with better self-reported general physical health. In addition, the correlation between the variable of self-reported general physical health and the variable of depression[3] was significant, r_s (520) $= -0.204$, $p = 0.001$, suggesting that more symptoms of depression were correlated with poorer self-reported general physical health. Although the statistics indicated that the strength of the correlation between mental and physical health status was small, the information derived from residents' opinions showed strong correlations between physical and mental health. Other than the statements listed in Table 5.2, four residents remarked on the effects of mood on both feelings and symptoms:

The emotional status is very important for physical health, especially for people living in big cities. Everybody faces lots of things in daily life, leading to stress. If you couldn't manage your mood and release the pressure in time, it will be harmful to your health. So it is important to learn how to control emotions, which could prevent potential diseases. Nowadays, I control my emotions consciously and I am used to the way of dealing with things gradually, so I am not feeling as tired as before. (LNT_FM8).

To me, keeping a good mood is important for keeping physical health. If I was in a bad mood, I felt uncomfortable everywhere inexplicably. Factors that influence my mood could affect my comfort degree directly, such as the weather, work, and interpersonal relationships. (LNT_FY7).

[3] The interval data derived from the CES-D scale was used in the Spearman's rank correlations analysis.

5.4 I Am Unhappy Living Here, but There is Nothing I Can Do. If not, I wouldn't Suffer from Diabetes. (GV_FO6)

5.4.1 If I Was Under Great Pressure, I Would Gain Much Weight Very Easily. With the Big Stress of Work, I Have no Time to Do Exercise and I Eat More Than Usual. (LNT_MY9)

Such opinions imply that moods, emotions, and the perception of happiness are important in determining and maintaining physical health. Poor mental health could cause physical illness in most cases. The evidence in the literature supports this association, and there is growing recognition, for instance, that depression is associated with cardiovascular disease [5], as cited in Carney and Freedland, 2009). The relations between positive emotions and physical health have been studied for more than a decade, and positive emotions are associated with fewer colds and less inflammation, and less likelihood of cardiovascular disease [6]. In addition, based on the analyses linking subjective well-being to objective health outcomes, [3] assert that subjective well-being could contribute to healthy functioning while subjective ill-being could result in compromised health. Moreover, different attempts have been made on the exploration of influential mechanism,for example, [4] argue that by affecting individual feelings, mental health and spiritual well-being could determine physical health,[6] point out that perceptions of social connections could promote positive emotions, which consequentially builds physical health. In short, the positive effects of mental health on physical health are supported by this study. In addition, mood was mentioned most frequently in terms of factors affecting physical health.

5.5 Conclusion

This chapter presents the assessment of residents' health in relation to physical and mental health. In general, it finds that the most residents in the three SCHEs are healthy, but with the underestimated prevalence of depression and level of stress.

In terms of physical health, majority residents were barely satisfied with their general physical health; while around one-third respondents were overweight, among which half of them were obese. In addition, the prevalence of obesity and general physical health were not significantly associated with the neighborhood type. With the similarities of physical health across the three SCHEs, this study proposes that the disparities of the physical environment of the three SCHEs have no influence on the physical health of residents, which will be explained in next chapter.

In terms of the mental health, 12.1% of respondents were diagnosed as depressed, which is much higher compared with other studies conducted in China. Risks of mental disorders were relatively low on the whole, and stress is more popular than anxiety and negatives effects of stress and anxiety on daily life among the residents. In addition, the prevalence of depression and the risks of mental disorders are different across the three SCHEs: AG has the most depressed residents; GV residents have fewer risks of mental disorders than the other two SCHEs. The association between mental health and neighborhood type will be further discussed in chapter 6, with the consideration of the other influential factors.

Moreover, physical and mental health interdepends on each other. Specifically, individual moods and emotions were the key influential factors for physical health, reflecting the crucial effects of mental health condition on physical health.

References

1. Cho MJ, Kim KH (1998) Use of the Center for Epidemiologic Studies Depression (CES-D) Scale in Korea. J Nerv Ment Dis 186(5):304
2. Halpern, D. (1995). *Mental health and the built environment: More than bricks and mortar?* Taylor & Francis US.
3. Howell RT, Kern ML, Lyubomirsky S (2007) Health benefits: Meta-analytically determining the impact of well-being on objective health outcomes. Health Psychol Rev 1(1):83–136. https://doi.org/10.1080/17437190701492486
4. Jackson RJ, Stacy S (2011) Designing Healthy Communities. John Wiley & Sons
5. Kemp AH, Quintana DS (2013) The relationship between mental and physical health: Insights from the study of heart rate variability. Int J Psychophysiol 89(3):288–296. https://doi.org/10.1016/j.ijpsycho.2013.06.018
6. Kok BE, Coffey KA, Cohn MA, Catalino LI, Vacharkulksemsuk T, Algoe SB, Brantley M, Fredrickson BL (2013) How Positive Emotions Build Physical Health: Perceived Positive Social Connections Account for the Upward Spiral Between Positive Emotions and Vagal Tone. Psychol Sci 24(7):1123–1132. https://doi.org/10.1177/0956797612470827
7. Lu Y, Hu P, Treiman DJ (2012) Migration and depressive symptoms in migrant-sending areas: Findings from the survey of internal migration and health in China. Int J Public Health 57(4):691–698. https://doi.org/10.1007/s00038-011-0314-0
8. Pan A, Lu L, Franco OH, Yu Z, Li H, Lin X (2009) Association between depressive symptoms and 25-hydroxyvitamin D in middle-aged and elderly Chinese. J Affect Disord 118(1–3):240–243
9. Radloff LS (1977) The CES-D Scale: A Self-Report Depression Scale for Research in the General Population. Appl Psychol Meas 1(3):385–401. https://doi.org/10.1177/014662167700100306
10. Roberts RE, Rhoades HM, Vernon SW (1990) Using the CES-D scale to screen for depression and anxiety: Effects of language and ethnic status. Psychiatry Res 31(1):69–83. https://doi.org/10.1016/0165-1781(90)90110-Q
11. Schmider E, Ziegler M, Danay E, Beyer L, Bühner M (2010) Is It Really Robust? Methodology 6(4):147–151. https://doi.org/10.1027/1614-2241/a000016
12. WHO. (2001). *The World Health Report 2001: Mental health: New understanding, new hope.*
13. Yang T (2002) Study on stress of urban residents during social transition. Chinese Epidemiology 23:473–475
14. Yang T, Huang L, Wu Z (2003) The application of Chinese health questionnaire for mental disorder screening in community settings in mainland China. Chinese Journal of Epidemiology 24(9):769–773
15. Yang T, Xu X, Li M, Rockett IRH, Zhu W, Ellison-Barnes A (2012) Mental Health Status and Related Characteristics of Chinese Male Rural-Urban Migrant Workers. Community Ment Health J 48(3):342–351. https://doi.org/10.1007/s10597-011-9395-8

Chapter 6
Factors Influencing Physical and Mental Health: Residential Environment and Health-Related Residential Behaviors

The literature has acknowledged that multiple factors derived from urban environments and lifestyles have crucial effects on individual's physical and mental health. Concerning urban transition and lifestyle changing, residential suburbanization and its effects on residents' health is a worthy subject to be explored. The risks of being obese and general physical health are inevitably in association with the way of life formulated by the living environment. Mental disorders and their causes in urban environment have been largely ignored in China's suburban living environment. Thus, this chapter presents influential factors in suburban living that are accountable for the prevalence of obesity, self-reported general physical health, the incidence of depression and risks of mental disorders (RMD) from the vantage points of residential environment and health-related residential behaviors. Regression analyses are used to identify the significant associations, followed by the explanation of the influential mechanism with the support of interviews, observations and literature.

Initially, this chapter demonstrates the necessity of controlling the variables of socio-demographic characteristics. Then the analysis is comprised of two main sections. The first section of analysis identifies the effects of residential environment and physical activity on residents' physical health, emphasizing the associations between physical activity and obesity, and between physical activity and self-reported general physical health. The quantitative analysis results are interpreted together with observation and interview findings to achieve a comprehensive understanding of the influence of the residential environment and physical activity on residents' physical health.

The second part of this chapter examines environmental and behavioral factors that are associated with depression and RMD. For environmental factors affecting the prevalence of depression, the objective factors associated with the physical residential environmental are the foci of the analysis considering the reported biased answers that might be given by people with the symptoms of depression. For environmental factors influencing RMD, variables of both physical and perceived residential environment are analyzed. For behavioral factors affecting the prevalence of depression,

T. Zhang, *Towards Healthy Settlements*, Urban Sustainability,
https://doi.org/10.1007/978-981-97-1207-6_6

attention is paid to the objective physical activity variables, while all three aspects of health-related residential behaviors (physical activity, interaction with neighbors, and residents' participation in housing management affairs) are examined for the influential factors for RMD. Finally, the factors influencing physical and mental health are summarized.

6.1 Socio-Demographic Characteristics of the SCHEs: The Control Variables

According to Chap. 2, socio-demographic characteristics are significant personal determinants of health, so they should be controlled in the exploration of the influential factors in regression analyses. Eight socio-demographic characteristics are controlled in the analysis of factors affecting health: age, gender, annual income, education status, employment status, family structure, homeownership, and length of residence.

Generally, socio-demographic features of the respondents are figured out based on the descriptive statistics. The average age was 39.63 years old ($SD = 14.90$); the mean average annual income was around 84,640 RMB ($SD = 121{,}081.32$); the gender distribution was nearly 1:1. Half (50.5%) of respondents were well-educated (with at least a junior college degree), and 38.9% were middle-educated, with a secondary or high school diploma. One-fourth (24.6%) of the respondents were unemployed or retired; for employed residents, half of them worked in Panyu district, 9.7% worked within the neighborhood and 10.7% in Tianhe district. For family structure, one-, two-, and three-generation families accounted for one-third of all households, respectively. Two-thirds (64.6%) of respondents were homeowners. The average length of residence was 4.60 year ($SD = 4.53$) (Table 6.1).

Comparatively, the annual income of the residents in the survey ($M = 84{,}640.47$, $SD = 121{,}081.32$) is higher than the average wage of employed people in urban districts of Guangzhou (73,133 yuan in 2014[1]). For educational status, slightly more than half of respondents (51.4%) were highly educated, which is much higher than the percentage of highly educated urban residents in Guangzhou in 2010 (19.22%).[2] In addition, the percentage of homeowners (64.6%) in this study is a little bit higher than the average (59.8%[3]) in overall population of Guangzhou, and the employed percentage (76.4%) of the respondents in this study is higher than that of urban

[1] Source: Guangzhou Statistical Yearbook 2015, retrieved from http://data.gzstats.gov.cn/gzStat1/chaxun/njsj.jsp.

[2] Source: The six census of Guangzhou (2010), retrieved from http://www.gzstats.gov.cn/tjgb/glpcgb/201105/t20110517_25227.htm.

[3] Housing conditions of urban households at year-end (2014), Guangzhou city statistical yearbook.

Table 6.1 Socio-demographic characteristics of total samples

	N	*M*	SD	Skewness		Kurtosis	
				Statistic	S.E	Statistic	S.E
Age	527	39.63	14.90	0.805	0.106	−0.238	0.212
Annual income (RMB)	428	84,640.47	121,081.32	5.638	0.118	50.516	0.235
Length of residents (year)	506	4.60	4.53	1.977	0.109	3.968	0.217

		N	Proportion
Sex	Male	258	49.0
	Female	259	49.1
	Total	517	100.0
Education status	Primary school and below	55	10.5
	Secondary school	84	16.0
	High school	120	22.9
	Junior college	127	24.2
	Bachelor or above	138	26.3
	Total	524	100.0
Employment status[4]	No jobs	129	24.6
	Within the neighborhood	51	9.7
	Panyu district	201	38.3
	Tianhe district	56	10.7
	Others	88	16.8
	Total	525	100.0
Family structure	One generation	178	33.9
	Two-generation	159	30.3
	Three-generation	158	30.1
	Others	30	5.7
	Total	525	100.0
Homeownership	Yes	337	64.6
	No	185	35.4
	Total	525	100.0

residents in Guangzhou (60.0%[5]). Therefore, compared with the overall population

[4] It could be biased due to the limitation that it is difficult to reach to the residents with long commuting needs in the field trip.

[5] The total number of employed persons at year-end is 7848,358, and the number of permanent population is 13,080,500. Source: http://data.gzstats.gov.cn/gzStat1/chaxun/njsj.jsp; http://www.gzstats.gov.cn/tjgb/qtgb/201504/t20150430_37572.htm.

of Guangzhou, the residents in the SCHEs are more in middle-age group, more-educated, with an above-average income, with high employment rate, and more are engaged in homeowner.

It is also important to know whether the socio-demographic characteristics of residents are different across the three SCHEs. One-way ANOVA and chi-square tests were conducted on the eight factors and the neighborhood type. Results showed that except for gender, all other socio-demographic characteristics varied across the three SCHEs significantly (Table 6.2).

GV residents ($M = 43.88$, $SD = 15.522$) were the oldest among the three SCHEs numerically, and AG were youngest ($M = 36.51$, $SD = 13.946$). One-way ANOVA was performed, suggesting that the mean difference was statistically significant, *Welch's F* (2, 326.370) = 10.560, $p < 0.001$(Table 6.2).

AG residents' annual income was highest both numerically and statistically, supported by the results of one-way ANOVA, *Welch's F* (2, 277.837) = 7.398, $p = 0.013$(Table 6.2).

In terms of the length of residence, residents in LNT ($M = 7.09$, $SD = 6.13$) had been longest in their neighborhood, followed by the residents in AG, and mean difference was also statistically significant, *Welch's F* (2, 274.674) = 60.522, $p < 0.001$(Table 6.2).

For education status, proportionally, more residents in AG (61.9%) than in the other two SCHEs had university degrees, and more residents in LNT (52.1%) than those in the other two SCHEs had graduated from high school. According to the chi-square test, the difference in proportion was statistically significant, X^2 (8, N = 524) = 47.685, $p < 0.001$(Table 6.2).

For employment status, more residents in GV did not work (40.7%) than those in the other two SCHEs, and fewer residents in GV worked within the neighborhood (2.1%) and in Panyu district (26.2%) than did residents of the other two SCHEs. Chi-square test indicated that the proportion difference was statistically significant, X^2 (8, N = 525) = 54.787, $p < 0.001$(Table 6.2).

The three SCHEs displayed different family structures: more residents in LNT (45.5%) were from one-generation families and more GV residents (48.6%) were from three-generation families. The proportions of different family structures were similar in AG. These proportion differences were statistically significant, X^2 (6, N = 525) = 44.362, $p < 0.001$(Table 6.2).

Finally, for homeownership, more residents in GV (79.7%) were homeowners. Only around half of the residents in LNT (51.4%) were homeowners, and this proportion difference was statistically significant, X^2(2, N = 522) = 27.391, $p < 0.001$(Table 6.2).

Thus, LNT residents were found to be the least educated and to have the highest proportion of those working near their home. The majorities of LNT residents were from one-generation families; and most were renters, not homeowners. The AG residents were the best educated and had the highest income. GV residents were the oldest and have the highest percentage of those who were retired, and the fewest residents worked within the neighborhoods, and the fewest residents being employed.

Table 6.2 Socio-demographic characteristics by SCHEs

	LNT	AG	GV
Age[a]	39.82 (14.66, 174)	36.51 (13.95, 208)	43.88 (15.52, 145)
Annual income (RMB)[b]	75,798.89 (146,863.15, 162)	103,548.73 (121,095.55, 158)	70,240.74(60,660.21, 108)
Length of residence[c]	7.09 (6.13, 166)	4.04 (3.38, 200)	2.46 (1.09, 140)
Sex[d]			
Male	52.5 (89)	47.0 (95)	51.4 (74)
Female	48.0 (82)	53.0 (107)	48.6 (70)
Total	171	202	144
Education status[e]			
Primary school and below	11.0 (19)	4.3 (9)	18.8 (27)
Secondary school	24.9 (43)	12.1 (25)	11.1 (16)
High school	27.2 (47)	21.7 (45)	19.4 (28)
Junior college	22.0 (38)	29.0 (60)	20.1 (29)
Bachelor or above	15.0 (26)	32.9 (68)	30.6 (44)
Total	173	207	144
Employment status[f]			
No jobs	17.8 (31)	18.9 (39)	40.7 (59)
Within the neighborhood	12.6 (22)	12.6 (26)	2.1 (3)
Panyu district	51.1 (89)	35.9 (74)	26.2 (38)
Tianhe district	6.9 (12)	12.1 (25)	13.1 (19)
Others	11.5 (20)	20.4 (42)	17.9 (26)
Total	174	206	145
Family structure[g]			
One generation	45.4 (79)	32.9 (68)	21.5 (31)
Two-generation	28.2 (49)	33.8 (70)	27.8 (40)
Three-generation	21.3 (37)	24.6 (51)	48.6 (70)
Others	5.2 (9)	8.7 (18)	2.1 (3)
Total	174	207	144
Homeownership[h]			
Yes	51.4 (89)	65.0 (134)	79.7 (114)

(continued)

Table 6.2 (continued)

	LNT	AG	GV
No	48.6 (84)	35.0 (72)	20.3 (29)
Total	173	206	143

Notes Percent (number of cases), mean (SD, N). * $p < 0.05$, ** $p < 0.01$, *** $p < 0.001$
For ANOVA, Welch's F has been reported when the assumption of homogeneity of variances has been violated
[a] *Welch's F* (2, 326.370) = 10.560^{***}, $p < 0.001$
[b] *Welch's F* (2, 277.837) = 7.398^{*}, $p = 0.013$
[c] *Welch's F* (2, 274.674) = 60.522^{***}, $p < 0.001$
[d] $X^2(2, N = 517) = 1.109$, $p = 0.574$
[e] $X^2(8, N = 524) = 47.685^{***}$, $p < 0.001$
[f] $X^2(8, N = 525) = 54.787^{***}$, $p < 0.001$
[g] $X^2(6, N = 525) = 44.362^{***}$, $p < 0.001$
[h] $X^2(2, N = 522) = 27.391^{***}$, $p < 0.001$

In addition, most GV residents were from three-generation families and were homeowners. It could therefore be concluded that the socio-demographic characteristics of the three SCHEs were quite different. So it is important to adjust the effects of them in the exploration of health determinants from residential environment and health-related residential behaviors. Prior to the regression analyses, the categorical variables are transformed into dummy variables, and the continuous data on age, income, and residential length are transferred into categorical ones (Table 6.3).

Table 6.3 Dummy variable list of socio-demographic variables

Variable	Dummy variable
Age	20–29 (0), 30–39, 40–49, 50–59, ≥60
Sex	Male (0), female
Income	<4w (0), 4–8w, 8–12w, >12w
Residential year	<1 year, 1–2 year, 2–3 year, 3–4 year, 4–5 year, ≥5 year (0)
Education status	Primary school and below, secondary, high school, junior college, bachelor or above (0)
Employment status	No jobs (0), have to work
Family structure	One generation family (0), two-generation family, three-generation family, others
Homeownership	Yes (0), no

6.2 Residential Environment and Physical Health

As discussed in Chap. 4, a methodological caveat for identifying the health determinants has been pointed out, suggesting the limits of the subjective measurements of residential environment in ascertaining the causes if residents' health. Constrained by the cross-sectional research, it is incapable to provide the consequential evidence to ascertain direct effects of the residential environment on residents' physical health.

As proved in Chap. 5, no differences were found across the three neighborhoods in obesity and self-reported general physical health. This indicates that the differences among the three types of physical residential environment have no influential effects on these two indicators of physical health. The research design and data of this study could only identify the physical environmental factors that are not associated with residents' physical health: the different characteristics of three SCHEs.

Due to the interdependent relationship between physical and mental health, the associations between perceived residential environment and self-reported physical health could be confounded by some mental disorders, which cannot be identified in this study. Thus, it is meaningless to study the relationships between these two subjective variables. Moreover, both literature and common sense suggest that the associations between perceived residential environment and prevalence of obesity are mediated by physical activity. So the effects of perceived residential environment on obesity is indirect, and the associations between them will be analyzed in Sect. 6.3.

Therefore, this study is limited in finding environmental factors associated with residents' physical health, but it could identify the physical environmental factors that have no significant impact on self-reported general physical health. The major differences of physical residential environment among the three SCHEs are analyzed based on data collected from observations, which may have no significant influences on obesity and general physical health condition (Table 6.4, Figs. 6.1, 6.2, 6.3, 6.4 and Appendix A).

For the transportation environment, the quality of walking surface in AG and GV is better than that in LNT, and walking in LNT is not as safe as walking in GV and AG because of the coexistence of pedestrian and vehicles in LNT (Table 6.4, Fig. 6.1). However, taking the advantages of land use pattern and street network, walkability of LNT is much better than that of the other two SCHEs (Table 6.4, Figs. 6.2, 6.3 and 6.4). In addition, LNT has the most convenient public transportation system with the most bus lines connecting to other regions. Although AG is at the disadvantage of no metro,[6] it is complemented by the shuttle bus provided by its developer (Table 6.4). Thus, the transportation environment of the three SCHEs has its own merits, and the significant differences among the transportation environment, such as pedestrian-vehicles relation, street density and morphology, and the public transportation system, may not have crucial effects on self-reported general physical health of the residents.

---

[6] The metro station near AG is under construction until the end of the filled trip of this study in 7th of June, 2016.

Table 6.4 Characteristics of residential environments of the three SCHEs

Domains		LNT	AG	GV
Transport environment	Walking surface	• Poor pavement • Sidewalk encroached by parking • Discrete shade tree	• Good pavement of sidewalk • Continuous shade tree and greening belt	• Good pavement of sidewalk • Continuous shade and greening belt • Walking trails
	Pedestrian-vehicles relationship	Coexist	Coexist and separate	Separate
	Public transit stops	1 subway station 8 bus stops	5 bus stops	1 subway station 14 bus stops
	Street density[7]	83.42 m/hm^2	24.57 m/hm^2	43.47 m/hm^2
	Street morphology (Average plot area)	Fine-grid pattern (5.54hm^2)	Curvilinear street pattern (8.60hm^2)	Grid pattern (10.93hm^2)
Leisure environment	Open space for leisure activity	• Public park • Riverside walkway	• Squares • Playing courts	• Squares • Playing court • Walking path • Urban greenway • Riverside walkway
	Fitness facilities and recreational center	Commercial fitness club	Commercial fitness clubs; Public recreational center	Public recreational center
Service environment	Health care facilities	Pharmacies (3); Community clinics (3[8]); Hospital	Pharmacies (1) Community clinics (1) Hospital	Pharmacies (1) Community clinics (1) Hospital
	Educational facilities	1 public primary schools 1 public middle school	3 public primary schools 1 public middle school; 1 private middle school	1 public primary school 1 public middle school

(continued)

[7] Street density = street length/site area, and streets within the residential clusters are excluded, because that they are only used by the residents in the same clusters.

[8] There are two dental clinics among the three community clinics.

Table 6.4 (continued)

Domains		LNT	AG	GV
	Daily commercial facilities	Active commercial activities	Moderate commercial activities	Inactive commercial activities
Environmental stress	Population density	93 units/hm^2 308 person/hm^2	36 units/hm^2 82 person/hm^2	43 units/hm^2 53 person/hm^2
	Green ratio	20%	36.8%	54.2%
	Ambient environment	Worst	Fair	Best
Design features	Identity features	None	• Reputation of the developer • Large-scale gated community	• Symbol of Asian Game (Tag and estates' name)
	Public realm	• No obvious public realm	• Lack of public realm at neighborhood scale • Well-designed open space in residential clusters	• Comprehensive planed and well-designed
	Aesthetic	• Hong Kong style	• Chinese style	• Lingnan (south of the Five Ridges) Style

For the leisure environment, the three SCHEs have similar types of opportunities for physical leisure activity, including public open and green space, gated open and green space (playing courts), public recreational center, and commercial fitness club (Table 6.4, Fig. 6.1). In terms of the quantity of the space and fitness facilities, Figs. 6.2, 6.3 and 6.4 show that the LNT has the least gated open and green space for physical leisure activity. In addition, the distribution of the public and gated open and green space in the three SCHEs is different. However, these differences seem have little impact on self-reported general physical health.

For the service environment, there is no significant difference among the three SCHEs in the quantity of educational facilities and health care facilities[9] (Table 6.4, Figs. 6.2, 6.3, and 6.4). However, the quality of the schools in GV is better than that of the other two SCHEs. For the daily commercial facilities, LNT has the most commercial space and most vigorous commercial streets, compared with the least daily commercial facilities in GV (Figs. 6.2, 6.3, and 6.4). Thus, it is argued that the

[9] Although there is no hospital in LNT, the hospital where LNT residents usually go could be accessed in 40 min by public transportation.

Fig. 6.1 Residential environments of the three SCHEs

differences in quality of schools and daily commercial services have no significant influence on self-reported general physical health.

For design features, the level of environmental stress of the three SCHEs is not high since all of them are located in suburban areas and enjoy relatively good ambient environment (Fig. 6.1, Appendix A). LNT has the lease identity design in terms of the sense of belonging; the aesthetic style across the three SCHEs is different. However, no effect of these differences on self-reported general physical health has been supported by this study.

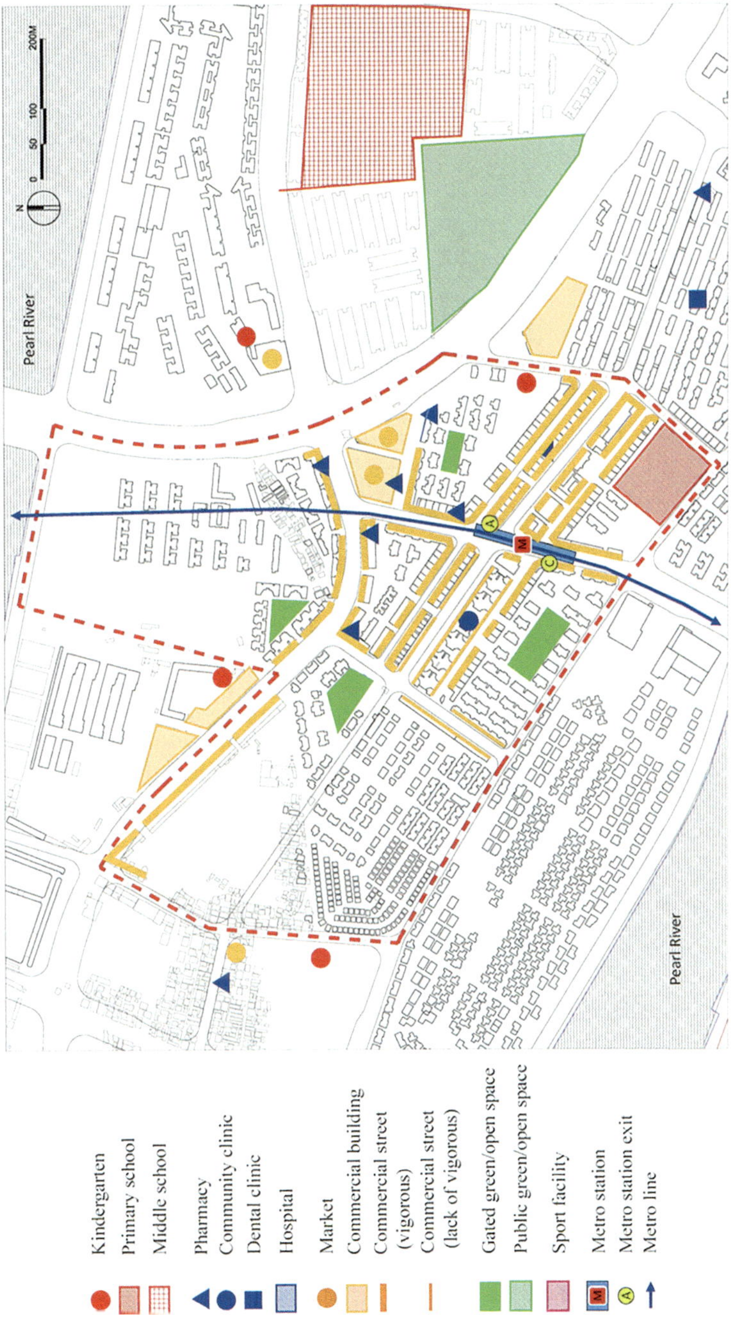

Fig. 6.2 Distribution of service facilities and green space in LNT

Fig. 6.3 Distribution of service facilities and green space in AG

Fig. 6.4 Distribution of service facilities and green space in GV

To summarize, this study points out that the differences among the three SCHEs in physical residential environment may have little impact on self-reported general physical health. These differences include: pedestrian-vehicles relation, street density and morphology, public transportation system, quantity of green space, distribution of the public and gated open and green space, quality of schools and daily commercial services, sense of belonging design and aesthetic style. The no effect of these differences on may lie on two possible reasons: these factors in three SCHEs have similar effects on self-reported general physical health; or these factors have no impact at all. Further study is needed to differentiate these two reasons.

6.3 Physical Activity and Physical Health

Major efforts are put on the effects of health-related residential behavior on physical health. Two hypotheses are put forward for exploring the influential factors: (1) physical activity pattern of living in SCHEs may have some effects on the incidence of obesity; (2) physical activity and mental disorders may have some effects on the self-reported general physical health. Regression analyses are used to identify the associations and the interviews are referred to for the causations afterwards.

6.3.1 Associations Between Physical Activity and Obesity

Binary logistic regression analysis was performed to study the associations between physical activity and the prevalence of obesity. The dependent variable was the prevalence of obesity, with being obese coded as 1 and being not obese coded as 0. Independent variables included both physical activity and socio-demographic variables. However, no variables of leisure activity were found associated with the odds of being obese in this study, so only one model was presented (Table 6.5).

Results showed that only walking time was significantly associated with the odds of being obese with the adjustment of travel mode for commuting, driving time per weekday and socio-demographics variables. The relative odds of being obese for residents walking more than 30 min was 0.000, indicating the little odds for residents walking more than 30 min being obese than the reference age group of less than 30 min ($p < 0.05$).

Statistical results suggested the significant association between walking and the odds of being obese, however, the directionality is still difficult to determine in a cross-sectional study. For example, in residential locations selection, obese persons may be less likely to give preference to the accessible physical activity and healthy foods compared with non-obese persons [38]. In addition, some studies argue that walking decreases with obesity and declines with weight gain, implying that the association between obesity and decreased walking might reflect greater body weight rather than a mechanist effect of weight gain [28]. To compensate for this limitation, many

Table 6.5 Results of logistic regression models examining association between odds of being obese and health-related residential behaviors

Model predictors	OR (95%CI) (N = 121)
Travel activity	
Travel mode for commuting	
Others and multiple choice—Reference	
Active travel mode	0.012 (0.000, 2.541) p = 0.106
Public transportation	0.002 (0.000, 1.769) p = 0.072
Private car	0.023 (0.000, 52.690) p = 0.340
Walking time per weekday	
<30 min—Reference	
≥30 min	0.000* (0.000, 0.424) p = 0.034
Driving time per weekday	
<0.5 h—Reference	
0.5–1 h	0.001 (0.000, 1.835) p = 0.072
>1 h	0.100 (0.000, 30.811) p = 0.431
Individual socio-demographics	
Age	
20–29—Reference	
30–39	5.403 (0.024, 1223.472) p = 0.542
40–49	1555.174 (0.426, 5,671,262.00) p = 0.079
50–59	46,092,161.5* (14.081, 1509E + 14) p = 0.021
≥60	2.644 (0.000, –) p = 1.000
Sex	
Female	0.000* (0.000, 0.179) p = 0.018
Income	
<4w—Reference	
4–8w	0.001(0.000, 3.044) p = 0.093
8–12w	0.010 (0.000, 23.257) p = 0.246
12w	0.381 (0.004, 40.069) p = 0.685
Residence length	
≥5 year—Reference	
<1 year	0.038 (0.000, 304.842) p = 0.476
1–2 year	0.012 (0.000, 3.192) p = 0.121
2–3 year	0.000 (0.000, –) p = 0.996
3–4 year	0.000* (0.000, 0.757) p = 0.042
4–5 year	0.051 (0.000, 24.545) p = 0.345
Education status	
Bachelor or above—Reference	

(continued)

Table 6.5 (continued)

Model predictors	OR (95%CI) (N = 121)
Primary school and below	4.478 (0.000, 5.733E + 187) p = 0.995
Secondary school	0.000 (0.000, –) p = 0.998
High school	0.001 (0.000, 3.258) p = 0.096
Junior college	60.720 (0.617, 5971.941) p = 0.079
Employment status	
No jobs—Reference	
Have to work	68,535.690 (0.000, –) p = 0.999
Family structure	
One generation—Reference	
Two-generation	0.011 (0.000, 6.591) p = 0.168
Three-generation	0.008 (0.000, 10.522) p = 0.186
Others	50.485 (0.000, –) p = 0.100
Homeownership	
No	0.002* (0.000,0.678) p = 0.036
Reliability	
R^2 (Cox & Snell)	0.293
R^2 (Nagelkerke)	0.673
−2log likelihood	27.129

studies transfer the focus to the relationships among physical activity, obesity, and residential environment, arguing that obesity is mainly affected by two behaviors, food consumption and physical activity, which are influenced by the environment [7, 32, 38]. The literature has established the associations between physical activity and obesity, and studies in both China and the West conclude that a higher level of physical activity could reduce the risk of obesity [4, 42]. For instance, a cross-sectional survey conducted in Tianjin city (China) found that "daily time on commuting or leisure-time physical activity was inversely related to mean BMI and prevalence of overweight among men" [24]. It is therefore accepted that a walkable environment could increase walking time, which is associated with obesity [17].

For control variables, statistical analysis suggested that age, gender, and residence length were associated with the odds of being obese. Age has a negative effect on health status, confirmed by the survey of self-reported health of population in Southern China [37]. In this study, in reference to age group 20–29, age group 50–59 has much higher odds of being obese in the adjustment model ($p < 0.05$) (Table 6.5). The relation makes sense since with increasing age, metabolism decreases, making weight control difficult and resulting in the high probability of becoming obese. In addition, female residents were much less likely to become obese than males ($p < 0.05$). Other than the biological difference, possible reasons may be that women are more weight-conscious than men and that self-perception of being overweight is more common in women than in men [40]. In the study, residents' attitudes towards the weight control indicates the difference between men and women (Table 6.6): females usually pay more attention to and spent more time on weight control than males, and females who do not try to control their weight are still aware of weight control. Women have more interest in weight control because of beauty concerns. Common reasons for not controlling weight include the effects of stress on diet and no time for exercise (Table 6.6. Thus, it is possible that male residents underestimate their weight status, have less interest in weight control, face more stress and are much busier at work than their female counterpart; these all lead to the higher percentage of obesity in men.

6.3.2 Associations Between Physical Activity and General Physical Health

To explore the influential factors of physical activity for the self-reported general physical health, a system hierarchy logistic regression analysis was performed. All of the models were with the same dependent variables, the self-reported general physical health, and this ordinal variable was transformed into a binary one, in which feeling good condition was coded as 1 and not feeling good condition (including fair and poor) is coded as 0. The independent variables of four models were different except the controlling variables: model 1 focused on the variables of travel activity; model 2 focused on the variables of physical leisure activity; model 3 included both variables of travel activity and physical leisure activity; model 4 involved the variable of risks of mental disorders to adjust the effects of mental health (Table 6.7).

Results in Table 6.7 shows that driving time per weekday and the preference for physical leisure activity were associated with the odds of being in good physical health across the models significantly and robustly, with the adjustment of other physical activity variables, RMD and socio-demographic variables. Detail associations are given in the following passages. For the control variables, age and family structure were significantly associated with the odds of being in good in general physical health robustly (Table 6.7).

The statistical results show that in reference to driving less than half an hour per weekday, driving more than one hour per weekday was associated with much higher odds of being in good general physical health, and the relative odds increased from adjusting for travel activity variables (Model 1), to adjusting for all physical activity variables (Model 3) and adjusting both physical activity and mental health variables (Model 4) (OD = 5.822, 95%CI = 1.240–27.327 in Model 1, OD = 9.727,95%CI = 1.521–62.226 in Model 3, OD = 16.364, 95%CI = 1.925–139.072 in Model 4). This association is unexpected, especially in light of studies that suggest negative relations between driving and health in the context of urban sprawl [18, 52]. To explain the counterintuitive associations between driving time and self-reported general physical health, two explanations or mediators are derived from interviews and literature.

Table 6.6 Comparison of attitudes towards weight control between female and male residents

	Statement	Weight control[10]	Reasons
Female			
1	*Of course I am controlling my weight, and I think all of the women should control their weight through the diet and taking exercise. It is for both health and beauty. (AG-FY4)*	✓	Health and beauty concerns
2	*I control my weight now, focusing on my physical health. (LNT-FM8)*	✓	Health concerns
3	*I pay attention to my body weight, and I usually eat less meat at dinner. (LNT-FM4)*	✓	Health concerns
4	*I am on a diet right now, and I keep doing physical exercise once a week and walking everyday for around half an hour. (LNT-FY7)*	✓	Health and beauty concerns
5	*I intended to control my weight, but it is very difficult for me. I am too busy to do physical exercise since I have to work in daytime and have to look after my child at night. For another, life is very stressful sometimes, which makes me eat more to relieve pressure. (LNT-FM2)*	○	Stress and time conflict
6	*I need to control my weight, but I haven't made it come true. It is difficult to stick with physical exercise in daily life. (LNT-FY1)*	○	Less of willpower
7	*I was a farmer when I was young, so I am not fat. (GV-FO4)*	✗	Good self-perception
Male			
1	*I do not need to control my weight, as I am thin. (GV-MM3)*	✗	Good self-perception
2	*My body weight has stayed the same all these years, so I don't need to control it. (GV-MO5)*	✗	Good self-perception
3	*I could not control my weight, and I gained 10 kg in this half a year since I was stressed. I am too busy and too tired to take exercise every day, if I have some time, I'd rather to go to bed. (LNT-MY9)*	✗	Stress and time conflict

(continued)

[10] ✓ refers to have controlled body weight, ✗ refers to didn't control body weight, ○ refers to didn't control body weight but have the consciousness to control.

Table 6.6 (continued)

	Statement	Weight control	Reasons
4	*I intended to control my weight from this February, making me feel goods. (AG-MY3)*	✓	Health concerns
5	*I stick with running every day. (LNT-MM6)*	✓	Health concerns

One possible reason is the suburban daily routine (suburban ways of life). It is different from that of central city due to the distinct land-use pattern, transportation, socioeconomic distinctions and social and cultural norms [13]. As stated in Chap. 3, moving to the suburbs is a lifestyle choice since more affluent families have specific living environment and lifestyle preferences. Focusing on the links between daily routine and health, a health-promoting lifestyle usually includes: self-actualization, health responsibility, nutrition awareness, physical fitness, interpersonal support, and stress management [56]. In China, it is a cultural recognition that car ownership is associated with affluence and success [4]. In this sense, this study proposes that some suburban residents choose driving when they travel, which may increase self-actualization through the representation of socioeconomic status. For example, one resident living in GV stated that:

> My son could not afford a car now, so he has to take the subway to work every day. He has to suffer the long distance to work, usually standing in the coach in the subway car. (GV_FO4)

This opinion suggested that owning a car was directly related to income, reflecting the socioeconomic status and limiting the choice of lifestyle.

In addition, there is evidence that commuting by automobile is linked with the stress and stress-related health problems. For some residents, driving is a more comfortable way of travelling, reducing their psychological distress and even improving their quality of life. Several residents remarked:

> It is very necessary to drive a car in daily life, or it will be very inconvenient. We have three cars in our family, and the main reason we need private cars is the convenience, which could enhance the quality of our life significantly. (AG_MO6)

> I would become very anxious if I don't have a car to drive. I am used to drive to work, which could help me to avoid taking the seriously crowded public transportation, keeping a better appearance when I arrive at my office. Sometimes, I can relax in my car even when I am on the way to the destinations with the music on and air conditioning on, and I could recharge myself. (GV_MY3)

Such opinions suggest that for some people, driving to their destinations was less stressful and more comfortable than using public transportation.

Residents who enjoy physical leisure activity were more likely to be in good general physical health, and this association was robust in models 2, 3, 4. In addition, the relative odds decreased from only adjusting for leisure activity variables (Model 2) to adjusting for all physical activity variables (Model 3) and adjusting for both

Table 6.7 Results of logistic regression models examining association between odds of being good in physical health and health-related residential behaviors

Model predictors	Model 1 OR (95%CI) (N = 118)[11]	Model 2[a] OR (95%CI) (N = 382)	Model 3 OR (95%CI) (N = 118)	Model 4 OR (95%CI) (N = 118)
Travel activity				
Travel mode for commuting				
Others and multiple choice—Reference				
Active travel mode	1.048 (0.115, 9.543) p = 0.967	–	0.841 (0.058, 12.134) p = 0.899	0.751 (0.045, 12.391) p = 0.841
Public transportation	2.453 (0.258, 23.333) p = 0.435	–	1.175 (0.085, 16.296) p = 0.904	1.103 (0.067, 18.283) p = 0.945
Private car	1.040 (0.116, 9.304) p = 0.972	–	0.705 (0.052, 9.447) p = 0.792	0.503 (0.032, 7.845) p = 0.624
Walking time per weekday				
< 30 min—Reference				
≥30 min	0.966 (0.274, 3.403) p = 0.957	–	0.857 (0.196, 3.753) p = 0.838	0.940 (0.202, 4.368) p = 0.938
Driving time per weekday				
< 0.5 h—Reference				
0.5-1 h	2.923 (0.589, 14.505) p = 0.189	–	2.390 (0.319, 17.916) p = 0.397	3.891 (0.390, 38.807) p = 0.247
> 1 h	5.822* (1.240, 27.327) p = 0.026	–	9.727* (1.521, 62.226) p = 0.016	16.364* (1.925, 139.072) p = 0.010
Willingness to walking: Compared to other means, I prefer to walk to destinations in daily life				
Agree—Reference				
Not sure	1.311 (0.148, 11.625) p = 0.808	–	0.565 (0.040, 7.939) p = 0.672	0.624 (0.045, 9.253) p = 0.745
Disagree	0.944 (0.194, 4.582) p = 0.943	–	0.294 (0.035, 2.493) p = 0.262	0.304 (0.033, 2.845) p = 0.297

(continued)

[11] The different sample sizes in different models in this table are caused by different number of missing data in each model.

Table 6.7 (continued)

Model predictors	Model 1 OR (95%CI) (N = 118)	Model 2[a] OR (95%CI) (N = 382)	Model 3 OR (95%CI) (N = 118)	Model 4 OR (95%CI) (N = 118)
Leisure activity				
Variety of leisure activities				
None—Reference				
One type	–	1.579 (0.858, 2.905) p = 0.142	0.413 (0.057, 3.007) p = 0.383	0.524 (0.067, 4.072) p = 0.537
Two types	–	1.227 (0.594, 2.536) p = 0.581	0.338 (0.033, 3.409) p = 0.357	0.329 (0.029, 3.680) p = 0.367
Preference for leisure activities: I love doing leisure activities such as physical exercise				
Agree—Reference				
Not sure	–	0.673 (0.295, 1.535) p = 0.346	0.526 (0.027, 10.184) p = 0.671	0.270 (0.011, 6.467) p = 0.419
Disagree	–	0.398* (0.177, 0.893) p = 0.025	0.038** (0.004, 0.369) p = 0.005	0.026** (0.002, 0.356) p = 0.006
Regularity of doing leisure activities: I did physical exercise regularly during past six months				
Agree—Reference				
Not sure	–	0.712 (0.363, 1.397) p = 0.323	0.569 (0.095, 3.415) p = 0.537	0.601 (0.096, 3.749) p = 0.586
Disagree	–	1.249 (0.608, 2.566) p = 0.546	0.647 (0.093, 4.511) p = 0.660	0.653 (0.089, 4.766) p = 0.674
Risks of mental disorders (RMD)	–	–	–	2.081 (0.813, 5.322) p = 0.126
Individual socio-demographics				
Age				
20–29—Reference				
30–39	0.695 (0.158, 3.047) p = 0.629	0.928 (0.471, 1.829) p = 0.830	0.282 (0.033, 2.413) p = 0.248	0.281 (0.027, 2.919) p = 0.288
40–49	0.062* (0.006, 0.644) p = 0.020	0.472 (0.203, 1.098) p = 0.081	0.014* (0.001, 0.292) p = 0.006	0.012* (0.000, 0.324) p = 0.009

(continued)

Table 6.7 (continued)

Model predictors	Model 1 OR (95%CI) (N = 118)	Model 2[a] OR (95%CI) (N = 382)	Model 3 OR (95%CI) (N = 118)	Model 4 OR (95%CI) (N = 118)
50–59	6,142,913.7 (0.000, –) p = 0.999	0.610 (0.215, 1.734) p = 0.354	226,235,480.8 (0.000, –) p = 0.999	9,007,169.39 (0.000, –) p = 0.999
≥60	6.834E + 9 (0.000, –) p = 0.999	0.503 (0.144, 1.756) p = 0.281	7.988E + 9 (0.000, –) p = 0.999	5.596E + 9 (0.000, –) p = 0.999
Sex				
Female	3.315 (0.772, 14.242) p = 0.107	1.349 (0.799, 2.562) p = 0.375	3.255 (0.439 24.102) p = 0.248	3.617 (0.414, 31.575) p = 0.245
Income				
< 4w—Reference				
4–8w	0.129 (0.013, 1.292) p = 0.082	1.585 (0.743, 3.380) p = 0.234	0.511 (0.037, 7.015) p = 0.615	0.817 (0.055, 12.110) p = 0.883
8–12w	0.164 (0.017, 1.624) p = 0.122	1.136 (0.506, 2.551) p = 0.758	0.341 (0.022, 5.340) p = 0.443	0.538 (0.028, 10.259) p = 0.680
12w	0.339 (0.043, 2.703) p = 0.307	0.969 (0.425, 2.158) p = 0.939	0.472 (0.044, 5.009) p = 0.536	0.679 (0.058, 7.961) p = 0.758
Residence length				
≥5 year—Reference				
< 1 year	0.424 (0.019, 9.234) p = 0.585	0.864 (0.279, 2.676) p = 0.800	1.165 (0.026, 52.037) p = 0.937	2.131 (0.030, 149.450) p = 0.727
1–2 year	0.126* (0.017, 0.961) p = 0.046	0.957 (0.428, 2.139) p = 0.915	0.128 (0.014, 1.187) p = 0.070	0.098 (0.010, 1.009) p = 0.051
2–3 year	0.272 (0.044, 1.675) p = 0.160	1.276 (0.589, 2.768) p = 0.536	0.291 (0.035, 2.426) p = 0.254	0.171 (0.016, 1.789) p = 0.140
3–4 year	0.251 (0.049, 1.278) p = 0.096	0.779 (0.386, 2.287) p = 0.485	0.186 (0.029, 1.167) p = 0.073	0.126* (0.016, 0.974) p = 0.047
4–5 year	0.271 (0.022, 3.328) p = 0.307	0.932 (0.380, 2.287) p = 0.878	0.171 (0.009, 3.420) p = 0.248	0.134 (0.005, 3.414) p = 0.223

(continued)

Table 6.7 (continued)

Model predictors	Model 1 OR (95%CI) (N = 118)	Model 2[a] OR (95%CI) (N = 382)	Model 3 OR (95%CI) (N = 118)	Model 4 OR (95%CI) (N = 118)
Education status				
Bachelor's degree or above—Reference				
Primary school and below	400,492,525 (0.000, –) p = 0.999	1.826 (0.537, 6.209) p = 0.335	265,704,268 (0.000, –) p = 0.999	75,427,806.1 (0.000, –) p = 0.999
Secondary school	3.368E + 9 (0.000, –) p = 0.999	1.000 (0.413, 2.423) p = 1.000	6.017E + 9 (0.000, –) p = 0.999	6.711E + 9 (0.000, –) p = 0.999
High school	5.782 (0.790, 42.334) p = 0.084	1.745 (0.796, 3.823) p = 0.164	6.611 (0.483, 90.392) p = 0.157	11.579 (0.706, 189.880) p = 0.086
Junior college	3.300 (0.751, 14.513) p = 0.114	1.037 (0.520, 2.068) p = 0.919	2.052 (0.337, 12.484) p = 0.435	2.731 (0.369, 20.195) p = 0.325
Employment status				
No jobs—Reference				
Have to work	0.000 (0.000, –) p = 0.999	1.037 (0.419, 2.725) p = 0.890	0.000 (0.000, –) p = 0.999	0.000 (0.000,–) p = 0.999
Family structure				
One generation—Reference				
Two-generation	3.611 (0.805, 16.200) p = 0.094	2.104* (1.099, 4.027) p = 0.025	8.053* (1.246, 52.063) p = 0.028	8.932* (1.268, 62.909) p = 0.028
Three-generation	0.985 (0.207, 4.691) p = 0.985	1.003 (0.481, 2.093) p = 0.994	2.488 (0.325, 19.041) p = 0.380	3.168 (0.372, 27.021) p = 0.292
Others	52,499,652.5 (0.000, –) p = 1.000	0.715 (0.244, 2.096) p = 0.540	53,654,616.8 (0.000, –) p = 0.100	78,890,309.7 (0.000, –) p = 1.000
Homeownership				
No	0.309 (0.062, 1.530) p = 0.150	1.349 (0.799, 2.276) p = 0.375	0.301 (0.048, 1.892) p = 0.201	0.404 (0.060, 2.726) p = 0.352
Reliability				
R^2 (Cox & Snell)	0.273	0.082	0.343	0.357
R^2 (Nagelkerke)	0.414	0.119	0.521	0.542
-2log likelihood	89.368	408.766	77.326	74.77

physical activity and mental health variables (Model 4) (OD = 0.398, 95%CI = 0.177–0.893 in Model 2, OD = 0.038, 95%CI = 0.004–0.369 in Model 3, OD = 0.026, 95%CI = 0.002–0.356 in Model 4). This relation has been reviewed in Chap. 2 already, pointing out that physical activity, especially active leisure activity, could prevent or reduce chronic and heart disease, obesity, and mental diseases [15, 16, 26, 39]. Since people who enjoy physical leisure activity engage in more leisure activity, it is reasonable that they would be more likely to report good physical health.

To summarize, this study links three factors with residents' physical health: walking time per weekday, driving time per weekday, and preference for leisure activity. In general, it seems that the physical activity pattern of residents in SCHEs was not adverse to their physical health greatly, especially for the effects of driving activity. Thus, with the known negative effects of physical inactivity and sedentary lifestyle on physical health, it is open to question whether the residents of the SCHEs in Guangzhou are sedentary, car-dependent and physically inactive. Thus, based on data from questionnaire survey, the next section reveals patterns of physical activity among SCHEs' residents.

6.3.3 Sedentary Lifestyle or Not: Physical Activity Pattern in SCHEs

According to the empirical review in Chap. 3, Chinese urban residents are more sedentary than their rural counterparts because of differences in daily routine (work activity, transportation, and leisure activity) [41]. However, it is hard to determine whether or not the suburban lifestyle could be characterized as sedentary.

6.3.3.1 Travel Activity

Travel includes all trips from one destination to the other, and the patterns of residents in the three SCHEs are examined in terms of mode of travel and travel time on a typical weekday. Overall, more than one-third of residents (38.3%) took active travel modes (walking and bicycling) and 36.7% took public transportation to work; 17.2% residents drove and 4.7% residents used multiple travel modes to work (Fig. 6.5).[12] To the overall responses, each resident had an average of 1.15 choices. The largest percentage (27.5%) walked to work, 24.7% took the metro and 23.1% rode the bus (Fig. 6.6). Thus, most residents chose active travel mode and public transport to go to work, and less than 20% of residents used private cars for commuting.

Across the three SCHEs, more residents in LNT (36.3%) walked to work or took the bus (28.1%) to commute, more residents in AG used private car (25.3%) and shuttle bus (15.4%) to commute, and more residents in GV used metro (42.0%) to commute (Fig. 6.7). Thus, it could be inferred that LNT could encourage walking and

[12] Other ways for travel are the taxi, shuttle bus, and motor-taxi.

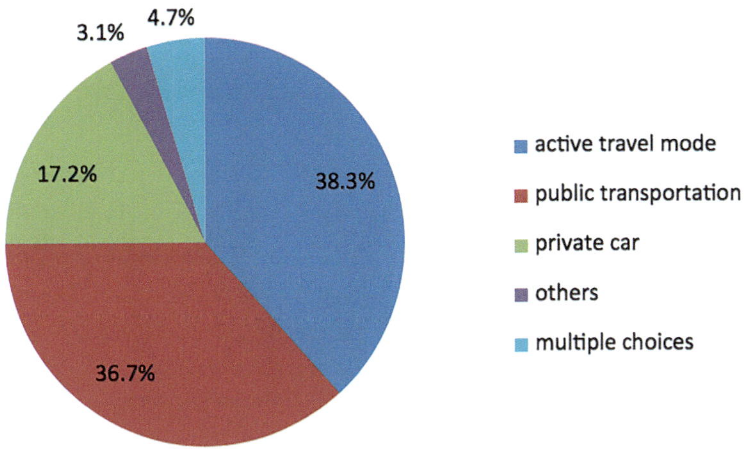

Fig. 6.5 Travel mode of commuting (N = 384)

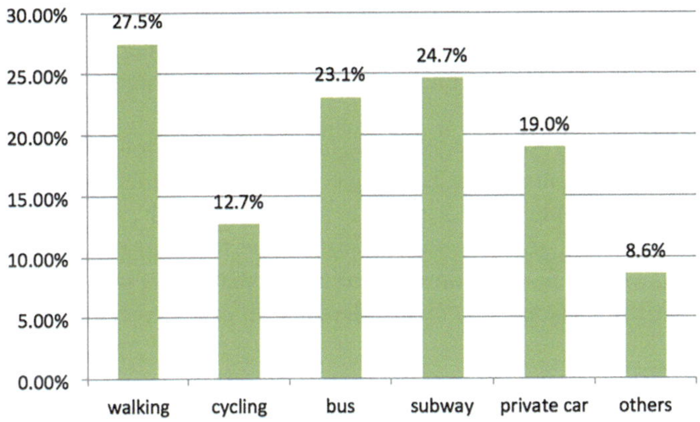

Fig. 6.6 Travel mode of commuting (Response = 445)

public transportation, while there are fewer jobs within walking distance, especially for GV residents. In addition, the shortage of public transportation service in AG was offset by the shuttle bus arranged by the housing management office; in this sense, there is little difference among the percentages of commuters taking public transport across the three SCHEs. For private car, only one-quarter of residents in AG drive to work. Therefore, it is hard to conclude the residents of the SCHEs are dependent on their cars; it is even possible that LNT could promote active travel.

In terms of the time spent on different travel modes, more than half of the respondents (52.2%) walked less than 30 min per day (Fig. 6.8). More than two-thirds (69.7%) took public transportation, and similar proportions (24.9% and 25.7%) spent less than 30 min and 30–59 min on public transportation, respectively (Fig. 6.9). More than half of the residents (58.6%) did not travel by private car. The largest percentage

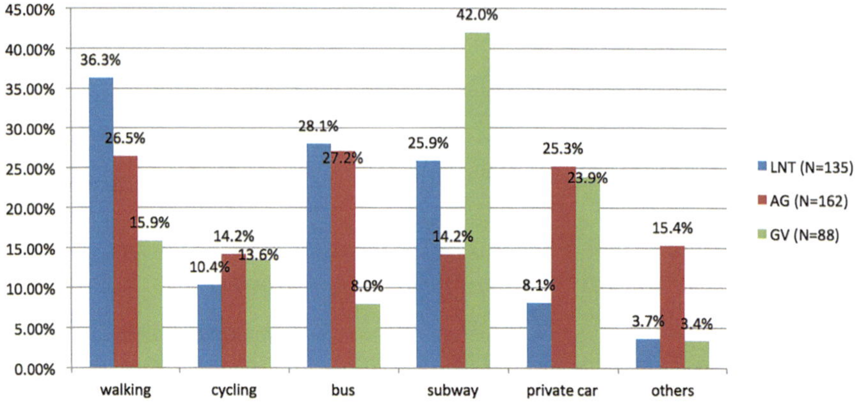

Fig. 6.7 Travel mode of commuting across the SCHEs (Response = 445)

(13.4%) drove less than 30 min, followed by driving for 30–59 min (10.1%) and more than 2 h (8.5%) (Fig. 6.10). For the overall travel time per weekday, around one half (53.6%) respondents spent less than two hours, while the other half (46.6%) spent more than two hours. This means the long distance from the designations for suburban residents (Fig. 6.11).

In conclusion, most residents took active travel and public transportation to work, and spent much time commuting. The long commutes could exacerbate the negative effects of travel on people's health, since an increase in health problems (back pain, cardiovascular disease, self-reported stress) is expected with more time spent on more crowded roads [18]. In addition, the different patterns of travel activity across the three SCHEs may stem from their differences in developmental stage, land use

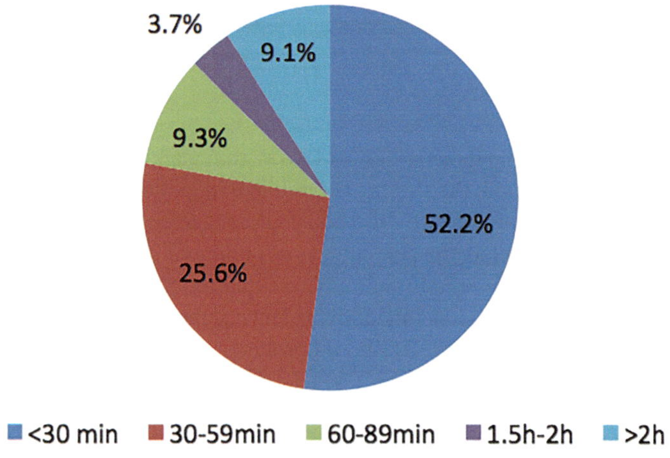

Fig. 6.8 Walking time per weekday (N = 429)

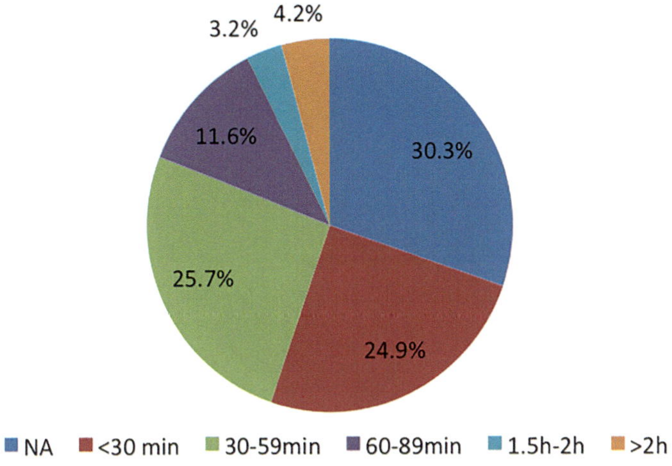

Fig. 6.9 Time spent on public transportation per weekday (N = 485)

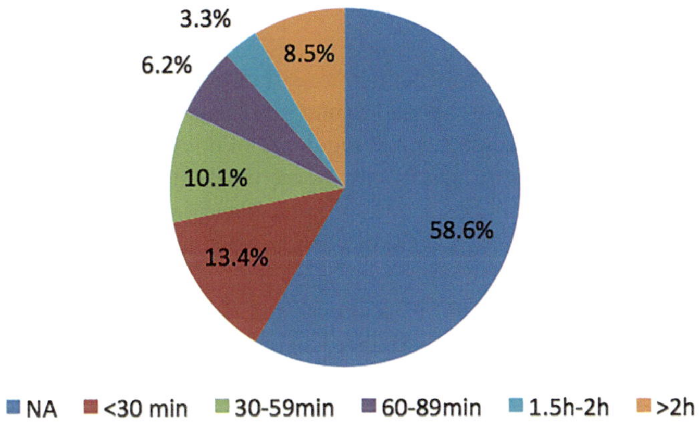

Fig. 6.10 Driving time per weekday (N = 498)

pattern and public transportation system as demonstrated in Chap. 3. More LNT residents used active travel and public transportation than private car. For private car use, although the percentages in AG and GV were both higher than that of LNT, the absolute percentages were not as high, implying that residents in AG and GV did not depend as much private cars for daily travel. Thus, it is argued that the mixed land use in LNT contributes to the active travel (walking and cycling) and provides jobs near homes, while the single land use pattern in AG and GV has failed in promoting active travel (walking and cycling).

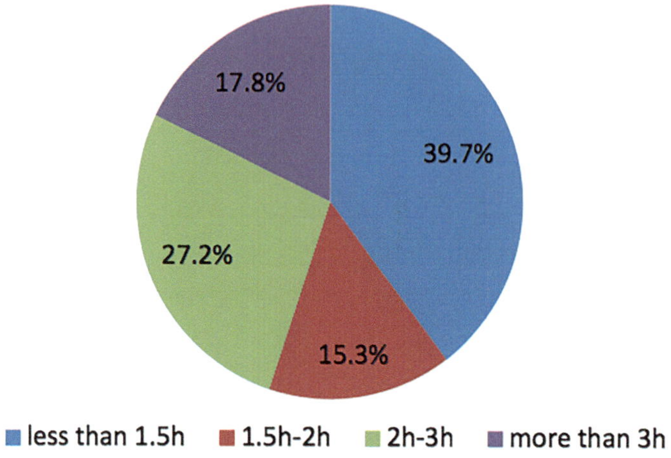

Fig. 6.11 Overall travel time per weekday (N = 478)

6.3.3.2 Physical Leisure Activity

For physical leisure activity, two aspects were examined through five indicators: (1) actual behaviors: variety, frequency and duration of leisure activities; (2) perception on doing leisure activity: preference and regularity of doing leisure activities.

In terms of variety of leisure activity, nearly one-third (31.4%) respondents did not participate in any leisure activities, while around half respondents (48.4%) conducted only one type of leisure activity in daily life, and around one-fifth (21.4%) joined two types. Nine types of leisure activities were mentioned, and the largest percentage of residents engaged in walking (24.5%) and playing ball games (17.6%), followed by jogging, swimming, dancing, doing fitness, yoga, and Tai Ji (Fig. 6.12).

Among the respondents who engaged in physical leisure activity, one-third (33.0%) did so every day, and more than half (56.7%) did more than three times per week (Fig. 6.13). One-fourth (25.1%) of the residents did leisure activities less than 30 min each time, less than half (44.8%) reported 30–60 min, and less than one-third of residents (30.1%) reported more than one hour each time (Fig. 6.14). In addition, across the three SCHEs, the results of chi-square test suggested that the actual behavioral patterns of physical leisure activity across the three SCHEs were similar in variety, frequency, and duration.

For the perception on doing physical leisure activity, most (73.2%) respondents expressed a preference for daily physical leisure activities, while only around half (50.2%) did physical leisure activities regularly in the previous six months. Across the three SCHEs, there was no difference in preference for leisure activity; however, the differences of perceived regularity of engaging in physical leisure activities was significant, X^2 (8, N = 518) = 37.255, $p < 0.001$. More residents in GV (58.1%) reported significantly more regular physical activities than residents in the other two

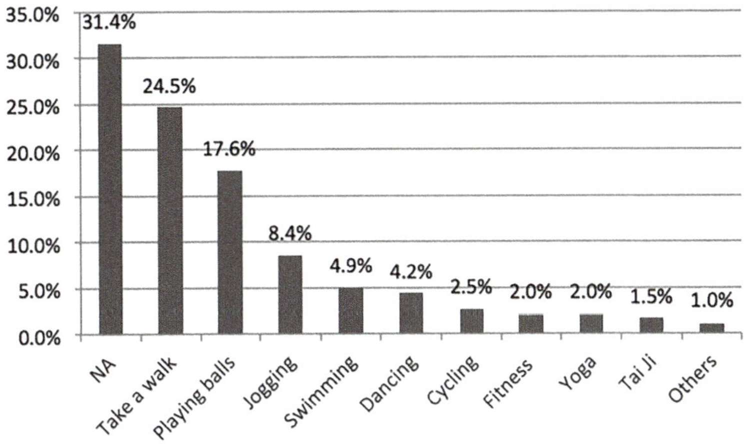

Fig. 6.12 Types of physical leisure activity (N = 527)

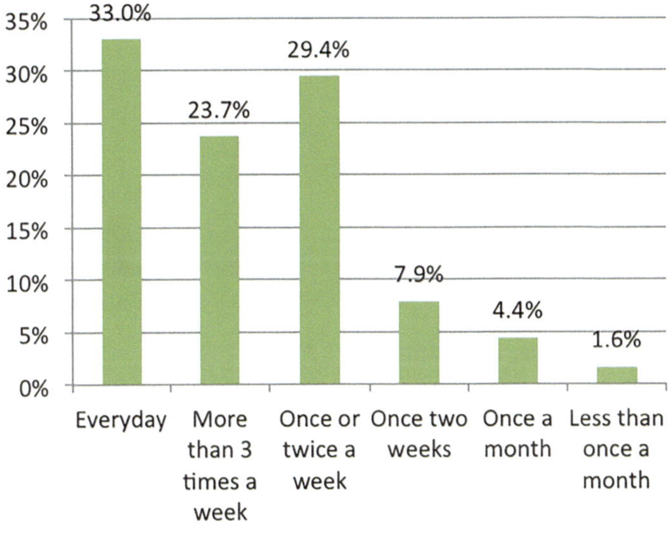

Fig. 6.13 Frequency of doing leisure activity (N = 367)

SCHEs, while more residents of LNT (37.0%) than that of the other SCHEs asserted that they did not take regular leisure activities (Fig. 6.15).

Thus, a majority of respondents preferred physical leisure activities, with two-thirds of them engaging in one type and one-third engaging in two or more. Among the residents who took leisure activities, more than half of them did the activities more than three times a week, and this was consistent with the number of residents who reported regular leisure activities. While the duration of the leisure activities for each time was moderate, three-quarters of the respondents (74.9%) spent more

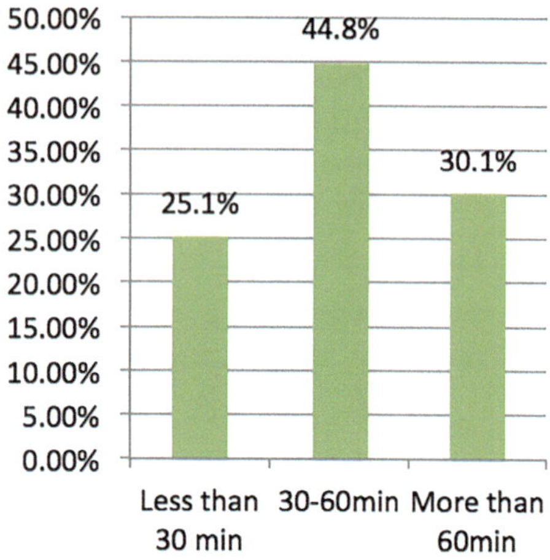

Fig. 6.14 Duration of leisure activity (N = 367)

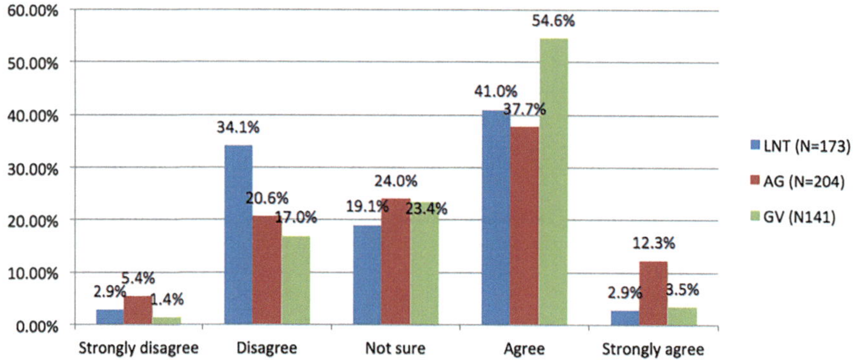

Fig. 6.15 Regularity of doing leisure activity (N = 518)

than 30 min on them (Table 6.8). Across the three SCHEs, there was no difference in terms of the conducted physical leisure activity; only the perceived regularity of doing physical leisure activity was significantly different.

These findings are varied from those in the literature. For example, the study conducted in Tianjin showed that more than two-thirds of urban residents did not engage in physical leisure activity at all [24], while this study finds that 79.8% did. For travel activity, Hu et al. [24] point out that 40% of urban residents in Tianjin spent at least 30 min commuting, and this study finds that the 38.3% did. Overall, in travel activity and physical leisure activity, there was not enough evidence to conclude that living in SCHEs made residents sedentary or physically inactive. However, although the proportions of using active travel mode and conducting leisure activity were relatively high, the amount of energy consumption and intensity of physical leisure activity may not be enough for keeping health, such as the relatively low proportion of residents walking more than 30 min, which is the recommended minimum criterion

Table 6.8 Leisure activity pattern

Indicators	Results
Variety of leisure activity	• No leisure activity: 31.4% • One type: 48.4% • Two or more types: 21.4%
Frequency of doing leisure activities	• Every day: 33.0% • More than 3 times a week: 56.7%
Duration of doing leisure activities	• Less than 30 min: 25.1% • 30–60 min: 44.8% • More than one hour: 30.1%
Preference for leisure activity	• 73.2% respondents expressed their preference
Regularity of physical leisure activity	• 50.2% did regular leisure activities

by WHO for maintaining healthy. In addition, the attendance of physical leisure activity may not be regular enough to keep a healthy lifestyle. Interviews could support these arguments:

> Although I play badminton frequently, the intensity is not enough. I usually have a little time to do physical exercise because of work. (AG_MY7)

> I am very busy with my work, and the most frequent exercise I take is walking. After a busy day, I do not want to do any vigorous exercise. (LNT_MY9)

These opinions indicate that the amount of exercise of most working population is far from enough due to the time limitation and working stress. Thus, this study argues that, generally, it could not rate the suburban living in SCHEs as sedentary lifestyle, while there is still far distance between suburban lifestyle and health-promoting lifestyle, especially for the intensity and regularity of doing physical activity.

6.4 Residential Environment and Mental Health

6.4.1 Associations Between Residential Environment and Depression

To identify the associations between residential environment and depression, binary logistic regression analysis was conducted with the dependent variable of depression (depressed was coded as 1 and not depressed is coded as 0) and independent variables of neighborhood type, housing type and eight socio-demographic variables (Table 6.9). To be clear, only objective measurements of residential environment were used in the regression analysis due to the potential reported biases caused by depressive disorders. Two models were performed with different reference groups of neighborhood type.

Table 6.9 Results of logistic regression models examining association between odds of being depressed and residential environment

Model predictors	OR (95%CI) (N = 215)	OR (95%CI) (N = 215)
Neighborhood type		
LNT	Reference	0.988 (0.082, 11.978) p = 0.993
AG	13.052** (2.284, 74.579) p = 0.004	12.899* (1.591, 104.599) p = 0.017
GV	1.012 (0.083, 12.264) p = 0.993	Reference
Housing types		
Around 30 floors—Reference		
Villa or townhouse	0.000 (0.000, –) p = 0.998	0.000 (0.000, –) p = 0.998
4–7 floors (without an elevator)	0.477 (0.065, 3.510) p = 0.467	0.477 (0.065, 3.510) p = 0.467
Around 11 floors	0.402 (0.055, 2.935) p = 0.369	0.402 (0.055, 2.935) p = 0.369
Around 17 floors	0.797 (0.120, 5.305) p = 0.814	0.797 (0.120, 5.305) p = 0.814
Individual socio-demographics		
Age		
20–29—Reference		
30–39	0.451 (0.096, 2.118) p = 0.313	0.451 (0.096, 2.118) p = 0.313
40–49	0.484 (0.079, 2.970) p = 0.433	0.484 (0.079, 2.970) p = 0.433
50–59	0.448 (0.040, 5.035) p = 0.516	0.448 (0.040, 5.035) p = 0.516
≥60	0.171 (0.007, 4.397) p = 0.287	0.171 (0.007, 4.397) p = 0.287
Sex		
Female	0.225* (0.061, 0.828) p = 0.025	0.225* (0.061, 0.828) p = 0.025
Income		
<4w—Reference		
4–8w	0.038* (0.003, 0.550) p = 0.016	0.038* (0.003, 0.550) p = 0.016
8–12w	0.658 (0.127, 3.411) p = 0.618	0.658 (0.127, 3.411) p = 0.618
12w	0.757 (0.127, 4.519) p = 0.760	0.757 (0.127, 4.519) p = 0.760
Residence length		
≥5 year—Reference		
<1 year	0.080* (0.008, 0.848) p = 0.036	0.080* (0.008, 0.848) p = 0.036
1–2 year	0.379 (0.052, 2.742) p = 0.337	0.379 (0.052, 2.742) p = 0.337
2–3 year	0.184 (0.017, 1.994) p = 0.164	0.184 (0.017, 1.994) p = 0.164
3–4 year	0.235 (0.042, 1.320) p = 0.100	0.235 (0.042, 1.320) p = 0.100
4–5 year	0.752 (0.099, 5.706) p = 0.782	0.752 (0.099, 5.706) p = 0.782
Education status		
Bachelor or above—Reference		

(continued)

Table 6.9 (continued)

Model predictors	OR (95%CI) (N = 215)	OR (95%CI) (N = 215)
Primary school and below	14.755 (0.828, 263.008) p = 0.067	14.755 (0.828, 263.008) p = 0.067
Secondary school	0.732 (0.056, 9.553) p = 0.812	0.732 (0.056, 9.553) p = 0.812
High school	3.325 (0.649, 17.038) p = 0.150	3.325 (0.649, 17.038) p = 0.150
Junior college	2.254 (0.437, 11.636) p = 0.332	2.254 (0.437, 11.636) p = 0.332
Employment status		
No jobs—Reference		
Have to work	2.600 (0.301, 22.460) p = 0.385	2.600 (0.301, 22.460) p = 0.385
Family structure		
One generation—Reference		
Two-generation	0.986 (0.262, 3.713) p = 0.983	0.986 (0.262, 3.713) p = 0.983
Three-generation	0.716 (0.117, 4.382) p = 0.718	0.716 (0.117, 4.382) p = 0.718
Others	1.200 (0.101, 14.249) p = 0.885	1.200 (0.101, 14.249) p = 0.885
Homeownership		
No	0.971 (0.216, 4.362) p = 0.970	0.971 (0.216, 4.362) p = 0.970
Reliability		
R^2 (Cox & Snell)	0.201	0.201
R^2 (Nagelkerke)	0.393	0.393
−2log likelihood	106.219	106.219

Results showed that in reference to the other two cases, living in AG was associated with higher odds of being depressed significantly adjusting for housing type and socio-demographic variables (Table 6.9). For control variables, sex, income, and residence length were associated with being depressed, adjusting for residential environment and other socio-demographic variables (Table 6.9). The odds of being depressed for AG was as high as 13.052 in reference to LNT residents ($p < 0.01$), (OR = 13.052, 95CI = 2.284–74.579 in Model 1). In reference to the group of GV residents, the odds of being depressed for AG residents was 12.899 ($p < 0.05$), (OR = 12.899, 95CI = 1.591–104.599 in Model 2). Thus, it is concluded that AG was associated with the greatest likelihood of being depressed.

The statistical results suggest that AG residents are the most likely to be depressed. This result seems to conflict with the common understanding, which could be explained by the difficulty of differentiating the two types of mechanisms for health determinants with limited evidence: selection (drifter mechanism) and causation (breeder mechanism) [11, 55]. Observation on the physical environmental characteristics of the three SCHEs could give some hints for understanding the influential mechanisms. Figure 6.1 and Table 6.1 have illustrated the characteristics of residential environment of the three SCHEs, implying that the residential environment of AG is not worse than that of the other two SCHEs but better than that of LNT especially in open and green space and fitness facilities (Appendix A).

Compared with the other two neighborhoods, AG has the medial residential density (36 units/hm^2), medial green ratio (36.8%), relative good walking surface, lowest street density (24.57 m/hm^2), diverse and abundant open spaces and fitness facilities for physical leisure activity, and good design features (such as fancy landscape and symbolic design), which could mean little ambient stress and positive design of the built environment for residents living in AG. Thus, from the perspective of three different types of physical residential environment, it is inferred that the association between neighborhood type and depression is due to the selection mechanism rather than causation. It is possible that the depressed individual or their family choose to live in AG where the residential environment could help releasing some symptoms of depression. However, without the longitudinal studies, this causation is still difficult to be ascertained.

6.4.2 Associations Between Residential Environment and RMD

In order to identify the factors of residential environment affecting RMD, a system of hierarchical multiple linear regression models have been conducted. Five models were performed with the same dependent variable, RMD, with the adjustment of the same eight socio-demographic variables (Table 6.10). For the independent variables, model 1 included two objective variables of residential environment, neighborhood type and housing type, which were both expressed in categories. The other four models included the variables measuring the four aspects of perceived residential environment (transportation, leisure and service environments and design features). To clarify, the five-point scale variables measuring residents' attitudes to the given statements were collapsed into agree, not sure, disagree. Preliminary analyses were performed to ensure that the assumption of normality, linearity and multicollinearity were satisfied.

Statistical results are presented in Table 6.10. Adjusting for socio-demographic variables, both neighborhood and housing type were significantly associated with RMD (model 1). The overall accessibility of destinations (public transport station, meat market, open space, schools, clinics, service center) within the SCHEs was associated with RMD significantly (model 2). Place chosen for physical leisure activity was associated with RMD significantly (model 3). Satisfaction with education facilities and daily commercial facilities were associated with RMD significantly (model 4). Quantity of green space and sense of belonging were associated with RMD significantly (model 5). For control variables, the significant associations between age and RMD were robust: in reference to age group 20–20, the age group 50–59 was associated with higher RMD score (except in model 3) (Table 6.10).

Table 6.10 Results of linear regression models examining association between RMD and residential environment

Model predictors	Model 1 β (95%CI) p (N = 366)[13]	Model 2 β (95%CI) p (N = 357)	Model 3 β (95%CI) p (N = 276)	Model 4 β (95%CI) p (N = 358)	Model 5 β (95%CI) p (N = 382)
Neighborhood type					
LNT—Reference					
AG	0.241* (0.014, 0.469) p = 0.037				
GV	0.127 (−0.152, 0.405) p = 0.372				
Housing types					
Villa or townhouse—Reference					
4–7 floors (without an elevator)	−0.166 (−0.563, 0.231) p = 0.411				
Around 11 floors	−0.261 (−0.671, 0.149) p = 0.211				
Around 17 floors	−0.478* (−0.922, −0.035) p = 0.035				
Around 30 floors	−0.215 (−0.716, 0.285) p = 0.398				
Perceived walking environment					
Accessibility to destinations within SCHEs		0.143* (0.005, 0.280) p = 0.042			
Quality of walking surface: The walking surface (pavement, street lamp, trees) is suitable for walking					
Agree—reference					

(continued)

[13] The different sample sizes in different models in this table are caused by different number of missing data in each model.

Table 6.10 (continued)

Model predictors	Model 1 β (95%CI) p (N = 366)	Model 2 β (95%CI) p (N = 357)	Model 3 β (95%CI) p (N = 276)	Model 4 β (95%CI) p (N = 358)	Model 5 β (95%CI) p (N = 382)
Not sure		−0.014 (−0.242, 0.215) p = 0.905			
Disagree		0.009 (−0.217, 0.235) p = 0.938			
Perceived leisure environment					
Place preference for leisure activity					
In neighborhood—Reference					
Outsider neighborhood			−0.401** (−0.685, −0.118) p = 0.006		
Others			0.036 (−0.240, 0.311) p = 0.800		
Perception of the quantity of open space: There is enough outdoor open space (squares, walking paths) to do exercise					
Agree—Reference					
Not sure			−0.077 (−0.355, 0.201) p = 0.586		
Disagree			−0.092 (−0.369, 0.186) p = 0.515		
Perception of the quantity of fitness facilities: There are not enough fitness facilities (playing courts, fitness equipment.) within the neighborhood					
Agree—Reference					
Not sure			−0.033 (−0.288, 0.222) p = 0.798		
Disagree			0.160 (−0.080, 0.400) p = 0.190		

(continued)

Table 6.10 (continued)

Model predictors	Model 1 β (95%CI) p (N = 366)	Model 2 β (95%CI) p (N = 357)	Model 3 β (95%CI) p (N = 276)	Model 4 β (95%CI) p (N = 358)	Model 5 β (95%CI) p (N = 382)
Satisfaction with quality of open space and fitness facilities					
Agree—Reference					
Not sure			−0.009 (−0.252, 0.234) p = 0.943		
Disagree			0.022 (−0.255, 0.300) p = 0.873		
Service environment					
Places chosen for health issues					
Community clinics—reference					
Hospital in Panyu				−0.065 (−0.271, 0.141) p = 0.535	
Hospital in urban core				0.001 (−0.223, 0.225) p = 0.991	
Others				0.324 (−0.070, 0.717) p = 0.107	
Satisfaction with education facilities: The schools within this neighborhood could not satisfy my demand in both quantity and quality					
Disagree—Reference					
Not sure				−0.122 (−0.325, 0.080) p = 0.235	
Agree				-0.269^{*} (−0.501, −0.037) p = 0.023	
Satisfaction with health care facilities: There are not enough health facilities for me					

(continued)

Table 6.10 (continued)

Model predictors	Model 1 β (95%CI) p (N = 366)	Model 2 β (95%CI) p (N = 357)	Model 3 β (95%CI) p (N = 276)	Model 4 β (95%CI) p (N = 358)	Model 5 β (95%CI) p (N = 382)
Agree—Reference					
Not sure				−0.008 (−0.284, 0.267) p = 0.953	
Disagree				−0.162 (−0.429, 0.105) p = 0.235	
Satisfaction with daily commercial facilities: There are not enough commercial facilities for me					
Agree—Reference					
Not sure				−0.112 (−0.403, 0.180) p = 0.451	
Disagree				0.339* (0.080, 0.599) p = 0.010	
Intangible environment					
Perception of safety: The neighborhood is safe from crime					
Agree—Reference					
Not sure					0.004 (−0.200, 0.207) p = 0.972
Disagree					−0.064 (−0.326, 0.198) p = 0.632
Perception of ambient environment: The ambient environment makes me comfortable (good air quality, good acoustic environment, good hygiene condition					
Agree—Reference					
Not sure					0.052 (−0.214, 0.317) p = 0.703

(continued)

Table 6.10 (continued)

Model predictors	Model 1 β (95%CI) p (N = 366)	Model 2 β (95%CI) p (N = 357)	Model 3 β (95%CI) p (N = 276)	Model 4 β (95%CI) p (N = 358)	Model 5 β (95%CI) p (N = 382)
Disagree					−0.037 (−0.270, 0.195) p = 0.753
Perceived density: I feel the population density is fine, neither too crowded or too deserted					
Agree—Reference					
Not sure					−0.031 (−0.234, 0.172) p = 0.762
Disagree					0.004 (−0.212, 0.220) p = 0.972
Perception of green space: The is not enough green space (trees and grass) in this neighborhood					
Disagree—Reference					
Other than disagree (Agree and not sure)					−0.200* (−0.369, −0.030) p = 0.021
Perception of public realm: There is enough social space within the neighborhood for neighbors' interactions and community activities					
Agree—Reference					
Not sure					0.019 (0.180, 0.219) p = 0.848
Disagree					0.039 (−0.173, 0.250) p = 0.720
Sense of aesthetic					
Agree—Reference					
Not sure					−0.113 (−0.300, 0.073) p = 0.232

(continued)

Table 6.10 (continued)

Model predictors	Model 1 β (95%CI) p (N = 366)	Model 2 β (95%CI) p (N = 357)	Model 3 β (95%CI) p (N = 276)	Model 4 β (95%CI) p (N = 358)	Model 5 β (95%CI) p (N = 382)
Disagree					−0.114 (−0.343, 0.114) p = 0.326
Sense of belonging: The design of this neighborhood gives me a feeling of attachment and belonging					
Agree—Reference					
Not sure					−0.194 (−0.420, 0.032) p = 0.092
Disagree					−0.387** (−0.675, −0.098) p = 0.009
Individual socio-demographics					
Age					
20–29—Reference					
30–39	0.009 (−0.206, 0.224) p = 0.936	0.038 (−0.182, 0.259) p = 0.732	−0.019 (−0.273, 0.236) p = 0.886	−0.014 (−0.226, 0.198) p = 0.898	−0.001 (−0.213,0.211) p = 0.991
40–49	0.134 (−0.148, 0.416) p = 0.350	0.153 (−0.137, 0.443) p = 0.300	0.117 (−0.218, 0.451) p = 0.493	0.107 (−0.226, 0.198) p = 0.457	0.109 (−0.171, 0.390) p = 0.444
50–59	0.519** (0.169, 0.870) p = 0.004	0.517** (0.161, 0.873) p = 0.005	0.395 (−0.004, 0.793) p = 0.052	0.430* (0.078, 0.782) p = 0.017	0.343* (0.005, 0.680) p = 0.047
≥60	0.269 (−0.161, 0.700) p = 0.219	0.153 (0–9.288, 0.593) p = 0.495	0.031(−0.421, 0.483) p = 0.891	0.077 (−0.355, 0.509) p = 0.726	−0.012 (−0.393, 0.369) p = 0.950
Sex					
Female	−0.075 (−0.242, 0.092) p = 0.377	0.091 (−0.260, 0.078) p = 0.292	−0.157 (−0.356, 0.042) p = 0.121	−0.133 (−0.301, 0.035) p = 0.120	−0.128 (−0.289, 0.033) p = 0.118
Income					

(continued)

Table 6.10 (continued)

Model predictors	Model 1 β (95%CI) p (N = 366)	Model 2 β (95%CI) p (N = 357)	Model 3 β (95%CI) p (N = 276)	Model 4 β (95%CI) p (N = 358)	Model 5 β (95%CI) p (N = 382)
<4w—Reference					
4–8w	0.000 (−0.239, 0.238) p = 0.997	0.015 (−0.226, 0.256) p = 0.903	0.063 (−0.212, 0.339) p = 0.651	−0.037 (−0.276, 0.201) p = 0.759	0.052 (−0.177, 0.281) p = 0.654
8–12w	0.047 (−0.220, 0.314) p = 0.728	0.021 (−0.252, 0.294) p = 0.878	0.052 (−0.258, 0.362) p = 0.740	0.013 (−0.256, 0.282) p = 0.924	0.074 (−0.183, 0.331) p = 0.571
12w	−0.044 (−0.317, 0.230) p = 0.753	−0.003 (−0.280, 0.273) p = 0.981	−0.117 (−0.433, 0.199) p = 0.466	0.020 (−0.253, 0.292) p = 0.886	−0.052 (−0.313, 0.208) p = 0.693
Residence length					
≥5 year—Reference					
< 1 year	0.100 (−0.267, 0.466) p = 0.593	0.222 (−0.140, 0.585) p = 0.229	0.098 (−0.305, 0.501) p = 0.633	0.300 (−0.056, 0.655) p = 0.099	0.077 (−0.260, 0.415) p = 0.652
1–2 year	0.277 (−0.011, 0.564) p = 0.060	0.361^{*} (0.087, 0.636) p = 0.010	0.233 (−0.078, 0.544) p = 0.141	0.354^{**} (0.088, 0.620) p = 0.009	0.196 (−0.067, 0.458) p = 0.143
2–3 year	0.180 (−0.092, 0.451) p = 0.194	0.270^{*} (0.015, 0.526) p = 0.038	0.212 (−0.069, 0.494) p = 0.139	0.330^{*} (0.078, 0.582) p = 0.010	0.181 (−0.054, 0.416) p = 0.131
3–4 year	0.112 (−0.141, 0.366) p = 0.384	0.148 (−0.091, 0.387) p = 0.223	0.201 (−0.070, 0.473) p = 0.146	0.254^{*} (0.016, 0.491) p = 0.036	0.138 (−0.091, 0.366) p = 0.237
4–5 year	0.186 (−0.134, 0.505) p = 0.254	0.202 (−0.105, 0.510) p = 0.196	0.120 (−0.235, 0.474) p = 0.507	0.234 (−0.067, 0.534) p = 0.127	0.074 (−0.222, 0.369) p = 0.623
Education status					
Bachelor or above—Reference					

(continued)

Table 6.10 (continued)

Model predictors	Model 1 β (95%CI) p (N = 366)	Model 2 β (95%CI) p (N = 357)	Model 3 β (95%CI) p (N = 276)	Model 4 β (95%CI) p (N = 358)	Model 5 β (95%CI) p (N = 382)
Primary school and below	−0.076 (−0.482, 0.330) p = 0.714	0.028 (−0.383, 0.439) p = 0.892	0.075 (−0.362, 0.513) p = 0.735	0.108 (−0.303, 0.518) p = 0.606	0.206 (−0.160, 0.573) p = 0.270
Secondary school	−0.238 (−0.544, 0.068) p = 0.126	−0.245 (−0.549, 0.058) p = 0.113	−0.139 (−0.485, 0.208) p = 0.431	−0.244 (−0.542, 0.055) p = 0.109	0.002 (−0.287, 0.290) p = 0.991
High school	−0.279* (−0.528, −0.021) p = 0.034	−0.215 (−0.474, 0.043) p = 0.103	−0.109 (−0.408, 0.190) p = 0.473	−0.165 (−0.426, 0.086) p = 0.196	−0.019 (−0.271, 0.233) p = 0.882
Junior college	−0.056 (−0.285, 0.173) p = 0.631	0.007 (−0.223, 0.237) p = 0.952	−0.047 (−0.325, 0.232) p = 0.743	0.006 (−0.218, 0.229) p = 0.961	0.025 (−0.209, 0.259) p = 0.833
Employment status					
No jobs—Reference					
Have to work	−0.184 (−0.507, 0.139) p = 0.262	−0.229 (−0.553, 0.096) p = 0.167	−0.229 (−0.568, 0.110) p = 0.185	−0.164 (−0.492, 0.163) p = 0.325	−0.270 (−0.555, 0.015) p = 0.064
Family structure					
One generation—Reference					
Two-generation	0.143 (−0.062, 0.348) p = 0.172	0.104 (−0.103, 0.312) p = 0.324	0.108 (−0.134, 0.350) p = 0.381	0.093 (−0.111, 0.296) p = 0.371	0.062 (−0.142, 0.267) p = 0.550
Three-generation	0.072 (−0.180, 0.323) p = 0.575	0.071 (−0.187, 0.328) p = 0.590	0.212 (−0.073, 0.497) p = 0.144	0.123 (−0.130, 0.377) p = 0.338	0.201 (−0.038, 0.440) p = 0.099
Others	−0.223 (−0.599, 0.153) p = 0.245	−0.145 (−0.518, 0.227) p = 0.444	−0.157 (−0.576, 0.262) p = 0.461	−0.115 (−0.499, 0.268) p = 0.555	−0.171 (−0.519, 0.177) p = 0.334
Homeownership					
No	−0.102 (−0.315, 0.110) p = 0.344	−0.178 (−0.387, 0.032) p = 0.096	−0.152 (−0.380, 0.076) p = 0.191	−0.156 (−0.362, 0.049) p = 0.135	−0.105 (−0.299, 0.088) p = 0.118

(continued)

Table 6.10 (continued)

Model predictors	Model 1 β (95%CI) p (N = 366)	Model 2 β (95%CI) p (N = 357)	Model 3 β (95%CI) p (N = 276)	Model 4 β (95%CI) p (N = 358)	Model 5 β (95%CI) p (N = 382)
R^2	0.122	0.111	0.197	0.163	0.209
F	F (28, 338) = 1.676 p = 0.019	F (25, 322)=1.659 p = 0.026	F (30, 246) = 2.006 p = 0.002	F (31, 327) = 0.2056 p = 0.001	F (35, 347) = 2.627 p < 0.001

6.4.2.1 Associations Between Physical Residential Environment and RMD

According to Model 1, in reference to living in LNT, AG was significantly associated with higher RMD score (referring to lower risks of mental disorders) with adjustment of housing type and socio-demographic variables ($\beta = 0.241$, 95%CI = 0.014–0.469) (Table 6.10). However, if GV is taken as the reference group, no significant association is found between the neighborhood type and RMD. Thus, the significant associations existed only when living in AG was compared to living in LNT. In this sense, there is no causal relationship between living in AG and higher prevalence of depression. The selection mechanism led to the observed relations.

In reference to living in villa or townhouse, living in a high-rise (around 17 floors) was significantly associated with lower RMD score, adjusting for neighborhood type and socio-demographic variables ($\beta = -0.478$, 95%CI = −0.922 to −0.035) (Model 1 in Table 6.10). This association is consistent with the literature: housing type does affect psychological health. People living in high-rise buildings may have more mental issues than those living in low-rise or houses [14]. Several mediators in the literature focus on the linkages between housing and psychological well-being: identity, insecurity, social support, and control [14]. It is believed that home is a place providing security, control, and reflecting identity through a symbol of self, representing inwardly and outwardly who we are, what we have achieved, and what we stand for [14]. Fail to reside in a place consonant with one's ideals therefore may affect self-esteem (Halpern [21]). Rosenbaum et al. [45] find that people relocating from public housing to middle-class suburban neighborhoods in the US felt increased self-efficacy and mastery than did people who relocated from public housing to a low-income neighborhood. Thus, this study proposes that a villa or townhouse may reflect a better living environment and higher social status, increasing the sense of identity, self-esteem and control over one's life and reducing the risks of mental disorders.

6.4.2.2 Associations Between Perceived Walking Environment and RMD

Model 2 showed that with each one-unit increase in accessibility of destinations within walking distance, the RMD score would increase 0.143, adjusting for quality of walking surface and socio-demographic variables, ($\beta = 0.143$, 95%CI = 0.005–0.280) (Table 6.10). This indicates that higher accessibility of destinations was associated with lower risks of mental disorders.

Evidence in literature shows supports for this association, in particular with the effects of physical activity as a mediator. Many studies confirm the association between walkability, connectivity and physical activity, especially walking [7, 44]. Related to walkability and connectivity, good accessibility of destinations in neighborhoods could promote more walking. In addition, much research has proved the positive effects of physical activity on mental health as discussed in Chap. 2. Thus, the association between accessibility of destinations and RMD could be mediated by the possibility of walking. The association between physical activity and RMD will be examined in Sect. 6.5.2.1.

6.4.2.3 Associations Between Perceived Leisure Environment on RMD

Model 3 suggested that the variable of place preference for physical leisure activity was associated with RMD with adjustment of other variables Table 6.10). In reference to doing leisure activity within the neighborhood, engaging in leisure activity outside the neighborhood was associated with lower RMD score with adjustment of other variables ($\boldsymbol{\beta} = -0.401, 95\%\text{CI} = -0.685$ to -0.118) (model 3 in Table 6.10). Thus, residents doing physical leisure activity in the SCHEs had fewer risks of mental disorders than those engaging in it outside the SCHEs.

It is inferred that attractive opportunities for physical leisure activity within the neighborhood could persuade residents to engage in physical leisure activity more frequently. Such activity could contribute to better mental health, reducing stress, increasing self-esteem and decreasing depression. In particular, the physical activity exposed to nature (green exercises) has more mental health benefits [39, 42]. Thus, the health benefits of physical activity could explain the association between place preference for physical leisure activity and RMD.

For the current opportunities for physical leisure activity, three residents noted:

> There is a lack of fitness and space for physical leisure activity in this neighborhood, such as the swimming pool and the courts for ball games. It is very inconvenient, and we have to go outside to do leisure activities. (LNT-FY1)

> We do not have large open space to conduct leisure activity, so we usually go outside to have a fun, going to the parks around the neighborhood. (LNT-FM4)

> There are some fitness facilities in the public space, but most of them are for children, and there were only stone seats and children's slide. In addition, the maintenance is not very good. Adults and the elderly usually go to the riverside to do some exercises. (LNT-FM5)

Most LNT residents unsatisfied with the opportunities for physical leisure activity due to the limited and substandard open space and fitness facilities. It can thus be concluded that: adequate, accessible and attractive opportunities within the neighborhood could encourage residents to engage in physical leisure activity more frequently. Thus, whether the SCHEs could provide attractive opportunities for residents to do leisure activity significantly influence residents' choice of where to engage in physical leisure activity.

6.4.2.4 Associations Between Perceived Service Environment and RMD

Model 4 suggested that satisfaction with education facilities and daily commercial facilities were significantly associated with RMD. In reference to satisfaction with education facilities, the dissatisfied group was associated with lower RMD score adjusting for other variables ($\beta = -0.269$, 95%CI = −0.501 to −0.037) (model 4 in Table 6.10). In reference to dissatisfaction with daily commercial facilities, the satisfaction group was associated with higher RMD score adjusting for other variables, ($\beta = 0.339$, 95%CI = 0.080–0.599) (Model 4 in Table 6.10).

For explaining these associations, few studies could be found due to the cross-culture issue, while a stress-distress model may be borrowed. This model highlights that social stressors, such as life events, particularly the undesirable ones, could exert moderate but steady effects on psychological distress [29].

For the choice of education facilities for children, it could be a significant social stressor for Chinese parents, especially for those living in SCHEs. The Chinese pay close attention to schools that their children will attend, contributing to extremely high price of housing in a school district. In reality, the educational resources that could be acquired by residents depend on the registered residence's location, which is usually the same with the housing location. Since there are fewer quality educational resources in suburban areas, residents in SCHEs have few choices.

In addition, before the housing commodification, work affiliation permeated everyday life including schooling. The responsibility of providing schooling facilities moved from the state to developers after housing commodification [29]. In the suburbs, developers have no initiative to provide good public schools, but they collaborate with private schools to make more profits (especially in the large-scale packaged SCHEs). Consequently, residents in SCHEs have to depend on the market for educational services, leading to families moving back to urban core area. However, due to the high housing price in urban core area, not all residents in the SCHEs could afford the movement. Thus, education is a life stressor for most parents with school-age children in the SCHEs. Interviews also supported this argument:

> The education issue gives me great pressure. The places in the kindergarten are very limit, there is no public kindergarten, and the price of private kindergarten is as high as 5000 RMD per month. In addition, the quality of education is worrisome. There are few choices for me even though I spend lots of money. (AG_MY7)

> After I gave birth to the baby, I cared more about the educational facilities around the neighborhood. There is only one public primary school, whose quality is just so-so. If there was a high-quality public school, I would be much happier. (AG-FY4)

> There are more private schools and kindergartens than public ones. If there were more public schools, that would be great. Our children need high quality public schools, which are affordable for most residents. The place in primary school is not enough, and I am considering whether we should move to another place to gain better educational resources, leading to a lot of pressure for me. (AG-FM1)

These opinions showed that choosing good schools is a major concern of young parents. With increased attentions being paid on children's education and constrained by the quality of the schools in SCHEs, it is reasonable to see schools as a social stressor, exposing residents in higher risks of mental disorders.

In addition, the association between daily commercial facilities and RMD is related to the stressor of economic status. In the interviews, several residents complained about either the over-priced goods in the markets (especially in AG) or the long distance of walking to shopping (especially in GV). Opinions imply that convenience and appropriate commercial type are important for the satisfaction with daily commercial facilities. Six residents noted that:

> Both the type and quantity of the commercial service could not satisfy our needs. We usually go shopping at urban central area. The commercial facilities here are relative low-end, and the turnover rate is quite high. (LNT_FY7)

> The price of the goods here is higher than that in urban central area. This could be due to the historical context that the first generation of residents here were rich. (LNT_MY8)

> I usually buy food in the meat market at the near village, because for the same goods, the price in the market here is much higher than that in the village's market. (AG-MO6)

> Because of the monopoly, the price of the goods in the market within the neighborhood is higher than that in the near village as much as 30–50%. (AG_MY7)

> My mom often asks me where she can buy more affordable vegetables. However, it is not convenient and safe for them to walk too long to shopping, which bother me to some extent. (AG_FM1)

> I have to walk to the area near the subway station to buy vegetables and meat. It is far away from my home, and I think it is not convenient. (GV-MO5)

> The commercial facilities are very bad. I have to go outside for having my hair cut. It is very inconvenient. In addition, I do not have any choice if I want to have a night snack in the neighborhood. (GV-MY3)

Such opinions indicate that in the three cases, residents were dissatisfied with the types and cost of the daily commercial service and therefore reported higher psychological distress. Thus, better educational and daily commercial facilities maybe protective factors against the risks of mental disorders by reducing the effects of social stressors. In particular, various choices of good schools, convenient and affordable shops could help to prevent the prevalence of mental disorders of residents.

6.4.2.5 Associations Between Perceived Intangible Environment and RMD

Model 5 suggested that insufficient green space and no sense of belonging were significantly associated with lower RMD score with adjustment of other variables, respectively, ($\beta = -0.200$, 95%CI $= -0.369$ to -0.030, $\beta = -0.387$, 95%CI $= -$ 0.675 t0 -0.098, respectively) (Table 6.10). Thus, it is ascertained that green space quantity and sense of belonging were associated significantly with RMD.

As discussed in conceptual background, the association between green space and mental health has been affirmed by numerous cross-cultural studies, such as the epidemiological studies conducted in England [33], Denmark [35], Sweden [6], and the Netherlands [19]. Increasing evidence shows that green landscapes have the capacity to alleviate mental fatigue, reduce stress and restore attention, while the lack of natural scenery can lead to increased aggression and violence [30, 53]. This is in line with the findings of [19], which confines very strong relationships between green space and anxiety disorders and depression. As well, based on three national, urban, and local cases, Maas et al. [30] point out that the quantity and quality of green space in residential areas are positively related to health.

To explain the causation, De Vries et al. [11] argue that more green space leads to better quality of the living environment, more frequent or longer contact with the green space, and more physical activity. Jackson [25] argues that the view of greenery and close proximity to green space are good for mental health through reducing stress and satisfying the human attraction to nature. In addition, Groenewegen et al. [19] point out that stress reduction, social cohesion and physical activity could explain the associations between green space and mental health, and the former two are more important explanatory factors. Thus, this study confirms that green space is a protective factor for risks of mental disorders. In the interviews, two residents noted that green space is essential for keeping calm and happy.

> I often feel less stressful when I come back and walk around the neighborhood, because there are much green space and wide-open space, and there are many residents seems relax. (AG-FM1)

> We do not have a yard, and there is no green space due to the open layout of housing blocks. We have to go to parks in other areas to enjoy the green space. We really need more green space here. (LNT-FM4)

In terms of the associations between sense of belonging and RMD, the literature offers several explanations. The sense of belonging is a basic human need or human motivation, and the need to belong fundamentally and pervasively motivates human beings to form and maintain everlasting interpersonal attachments [5, 31]. The sense of belonging can be a cultural element, related to psychosocial environment and the relationships among individuals and their community [23]. According to conceptual review, the absence of a sense of belonging could result in psychopathology [5, 20], while a strong sense of belonging is associated with more health-promoting behaviors [1].

To explain the above associations, social support and loneliness are often implicated, which could promote mental health and minimize mental disorders [20]. Hill [23] argues that the sense of belonging could promote health in communities as it created the sense of social support. For the elderly, social support and satisfaction with social contacts are significant protectors against depression [10]. In addition, the sense of belonging is a buffer against loneliness [2, 43], and loneliness has been proved to be linked with depression [20]. Thus, this study proposes that the residents' sense of belonging is associated with the risks of mental disorders as it gives rise to social support and the reduction of loneliness.

However, there is no consensus in the literature in terms of the causal links between the sense of belonging or social support and mental health [27, 47, 54]. The cross-sectional study is limited by the threat to validity, the potential bias in the retrospective recall of sense of belonging among the residents at higher risks of mental disorders [27].

In terms of the effects of living in SCHEs on sense of belonging, three residents remarked:

> The sense of belonging in suburban neighborhoods is stronger than in urban settlements. The relations between urban residents are much colder than they are among suburban residents. Especially, the elderly in suburban neighborhoods may feel less alone than their urban counterparts. (AG_MY7)

> I think the sense of belonging is not very strong in our neighborhood. In LNT, this sense could only be generated through the love of and familiarity of with physical environment, other than from the social atmosphere generated by the residents. You may not know your neighbors even though you have lived here for more than ten years. (LNT_FY7)

> Living in commodity housing estates can hardly generate the sense of belonging since most residents are from other cities. However, there are many activities in villages of Guangzhou, such as annual dragon-boat races and dragon-boat lunch [long chuan fan], through which strong sense of belonging could be generated and loneliness could be reduced. (LNT_MY8)

Such opinions suggest that the degree of the sense of belonging decreases from the rural to the suburban, then to the urban residential environment. Both physical environment and social networks affect the sense of belonging. Thus, it is inferred that the sense of belonging could be strengthened by the identity and symbolic meaning of the residential environment and by the social interactions and shared culture of the residents.

6.4.2.6 Robust Control Variables

Statistical results showed that age was robustly associated with RMD across the models in Table 6.4. To find out which age groups had the fewest RMD, one-way ANOVA analysis was conducted (Table 6.11). Numerically, the 20–29 group had the lowest average RMD score, the 30–39 and 40–49 groups had a similar average RMD score, and the 50–59 and ≥60 groups' average scores were the highest at more than 4(Table 6.11). The mean differences were statistically significant, $F\ (4, 521) = 9.876$, $p < 0.001$. The differences existed between 20–29 and 50–59, 20–29 and ≥60, 30–39 and 50–59, 30–39 and ≥60, 40–49 and 50–59, 40–49 and ≥60. The mean differences between the above groups were associated with moderate to large effect sizes based on the guidelines given by Haslam et al. [22] (Table 6.12). Therefore, residents aged from 50 to 59 were at the lowest risks of mental disorders, and the age 50 seems to be the threshold.

These associations between age and mental disorders are consistent with those of [9] who argues that young age is a risk factor for mental disorders. In controlled studies,[14] the patterns of the relationship consistently show a decrease of mental disorders across age groups [49]. Older people might be more adept at adjusting to the tough situation than young people. Older residents were also more likely

[14] The controlled factors for age and depression relationship are: "gender, race, marital status, education, employment status, income, household, place of residence, life event, social support, career role and somatic health status defined by the number of chronic illness present" [44, 48, 58].

Table 6.11 One-way ANOVA analysis of RMD by age group

	N	M	SD
20–29	136	3.80	0.79
30–39	175	3.88	0.77
40–49	76	3.87	0.79
50–59	64	4.41	0.78
≥60	75	4.26	0.86

Note The average RMD ranges from 1 to 5

Table 6.12 Results associated with Tukey's HSD multiple-comparisons

Comparison	*p*	*Cohen's d*[15]
20–29 to 50–59	<0.001	−0.78
20–29 to ≥60	0.001	−0.56
30–39 to 50–59	<0.001	−0.68
30–39 to ≥60	0.005	−0.47
40–49 to 50–59	0.001	−0.69
40–49 to ≥60	0.022	−0.47

financially secure and to have a more stable family life. Interviews revealed that the elderly loved discussing their contentment, especially if they had suffered when they were young. As two residents noted:

> I have three sons, and I suffered a lot raising them. Nowadays, they all have good jobs and life is getting much better, and I am basking in the love of my family. I am very happy. (AG_MO5)

> I am usually in very good mood, and there is nothing to worry about. My son is very filial pious and there is little economic pressure. (GV_MO5)

Thus, the emotional and economic stability were important in achieving a happy life, especially for older residents. In addition, rich life experiences could help to maintain emotional stability despite difficulties, thereby reducing the risks of mental disorders.

---

[15] According to Cohen's guideline, d = 0.2 is a small effect, d = 0.5 is a moderate effect, d = 0.8 or >is a large effect.

6.5 Health-Related Residential Behaviors and Mental Health

6.5.1 Associations Between Physical Activity and Depression

Bearing in mind the methodological caveat mentioned in Chap. 4, only the relations between physical activity and depression are examined, because depression is the reason of the reduction of interacting with neighbors and participating in housing management. Binary logistic regression analysis was conducted with the dependent variable of depression (depressed was coded as 1 and not depressed is coded as 0) and the independent variables of four physical activity variables and eight socio-demographic variables (Table 6.13).

Results showed that time on public transportation and variety of physical leisure activity were significantly associated with the odds of being depressed, with the adjustment of other variables (Table 6.13). In reference to spending less than 30 min on public transport per day, people who spent more than 30 min had 61.031 higher odds of being depressed ($p < 0.05$) (OR = 61.031, 95CI = 2.391–1558.071). In addition, people who participated in one type of physical leisure activity were 94% less likely to be depressed than the people in the reference group who participated in none ($p < 0.05$) (OR = 0.006, 95CI = 0.000–0.652) (Table 6.13). Thus, travelling by public transportation more than 30 min per day and engaging in no physical leisure activity was associated with much higher odds of being depressed.

Little research has been done on the effects of time spent on public transportation on depression. Time spent on public transportation could be the reason for depression, since it is not determined by the subjective perception. For the influential mechanism, interviews could give some hints. It was difficult to access and to interview the residents who have been diagnosed with depression, so residents who had not been diagnosed as depressed were interviewed about the negative effects of taking public transportation on their mood.

> I don't like to take public transport since the bus and taxi are very smelly and it takes time to wait. This would influence my mood. Usually I choose Uber to travel to my destinations, which is very convenient and comfortable. (LNT-FM2)

> The public transportation is very inconvenient, and the layout of bus line is not reasonable, with only a few choices to travel east of Panyu. In addition, the frequency is quite low, and the departure interval is as long as 25 min, leading to long waiting time, which usually makes me anxious and stressed. (AG_MY3)

> Of course, taking public transportation has negative impacts on my mood. It is very crowded in the coach in peak hours, and you do not need to hold the handrail at all since all around you are people. I almost could not stand this feeling. If I needed to take public transportation to work every day, I think I would be crazy. (LNT_MY7)

> It is terrible that you arrive at the station but you could not get on the bus or the subway unless you wait for another three or four. There are too many people in the peak hour. (AG_FY8)

Table 6.13 Results of logistic regression models examining associations between odds of being depressed and physical activity

Model predictors	OR (95%CI) (N = 117)
Physical activity	
Travel mode for commuting	
Multiple choice—Reference	
Active travel mode	0.940 (0.005, 167.016) p = 0.981
Public transportation	4.363 (0.042, 448.952) p = 0.533
Private car	19.089 (0.121, 3022.104) p = 0.254
Time spent on public transportation	
<30 min—Reference	
≥30 min	61.031*(2.391, 1558.071) p = 0.013
Variety of leisure activities	
None—Reference	
One type	0.006* (0.000, 0.652) p = 0.033
Two types	0.086 (0.002, 3.551) p = 0.196
Regularity of doing leisure activities: I did physical exercises regularly during past six months	
Agree—Reference	
Not sure	0.179 (0.002, 14.007) p = 0.439
Disagree	0.080 (0.002, 3.711) p = 0.197
Individual socio-demographics	
Age	
20–29—Reference	
30–39	0.005* (0.000, 0.445) p = 0.021
40–49	0.012* (0.000, 0.643) p = 0.029
50–59	1246.296* (3.813, 407,405.177) p = 0.016
≥60	1868.959* (1.400, 2,495,353.53) p = 0.040
Sex	
Female	0.007** (0.000, 0.294) p = 0.009
Income	
<4w—Reference	
4–8w	0.010* (0.000, 0.884) p = 0.044
8–12w	6.510 (0.273, 155.328) p = 0.247
12w	333.484* (2.179, 51,034.128) p = 0.024
Residence length	
≥5 year—Reference	
<1 year	0.090 (0.001, 6.045) p = 0.262
1–2 year	0.019* (0.000, 0.732) p = 0.033
2–3 year	0.000* (0.000,0.194) p = 0.028

(continued)

Table 6.13 (continued)

Model predictors	OR (95%CI) (N = 117)
3–4 year	0.023* (0.001, 0.684) p = 0.029
4–5 year	0.029 (0.001, 1.374) p = 0.072
Education status	
Bachelor or above—Reference	
Primary school and below	11.574 (0.009, 14,856.529) p = 0.502
Secondary school	0.568 (0.010, 31.431) p = 0.783
High school	1331.225** (6.917, 256,201.926) p = 0.007
Junior college	31.951* (1.063, 960.031) p = 0.046
Employment status	
No jobs—Reference	
Have to work	1.954 (0.052, 73.929) p = 0.718
Family structure	
One generation—Reference	
Two-generation	4.232 (0.277, 64.699) p = 0.300
Three-generation	0.039 (0.00, 2.544) p = 0.128
Others	0.807 (0.000, 19,138,444.5) p = 0.980
Homeownership	
No	0.212 (0.013, 3.340) p = 0.270
Reliability	
R^2 (Cox & Snell)	0.379
R^2 (Nagelkerke)	0.644
−2log likelihood	48.085

Such opinions suggest that the crowded coach, inconvenience of making transfer, and time-consuming waiting increase the travelers' anxiety and stress. Based on the physical characteristics of public transportation vehicles and perception of users, three explanations are proposed. According to the psychosocial stress explanation, which is similar to the stress-distress model, a poor quality of physical environment provokes psychosocial stress that might place people at greater risks for depression [29, 57]. The crowded passengers in peak hour reduce the quality of the vehicles environment, increasing travelers' psychosocial stress. In addition, the concentrated disadvantage explanation argues that: "the association may be a reflection of concentration of other deleterious elements that shape mental health" [57]. The long and uncontrollable transfer time and other passengers' impacts (the body odor and the train etiquette) could intensify the negative effects on moods, which in turn increase the risks of depression. Finally, the selection mechanism indicates that depressed people may have limited travel choices.

Moreover, previous studies have supported the association between a variety of physical leisure activity and depression. In other words, physical activity improves mood and reduces symptoms of depression [39, 46, 50]. Physical activity is an effective intervention for depression, which has been proved by experiments. It is found that depressed individuals receiving an aerobic-exercise intervention had more obvious improvements in depression comparable to those undergoing a psychotropic treatment [3]. Thus, although this study could not determine the causes and effects between the variety of physical leisure activity and prevalence of depression, it did find that based on the literature, physical leisure activity could intervene in the prevalence of depression.

6.5.2 *Associations Between Health-Related Residential Behaviors and RMD*

To identify the residential behavioral factors associated with RMD, multiple linear regression analyses have been performed, with the dependent variable of RMD, and with the three dimensions of health-related residential behaviors as independent variables, adjusting for socio-demographic variables. Preliminary analyses were performed to ensure that the assumption of normality, linearity and multicollinearity were satisfied. Since no factor of participation in housing affairs was found to be associated with RMD, only two models were presented in Table 6.14.

Results suggest that three variables of physical activity (walking time per day, total travel time per day, the regularity of doing leisure activity) were associated with RMD significantly with adjustment of other variables (Model 1 in Table 6.14). For neighbors' interaction, only the variety of informal neighbors' interaction was found to be associated with RMD significantly, adjusting other variables (Model 2 in Table 6.14).

6.5.2.1 Associations Between Physical Activity and RMD

Three physical activity factors were significantly associated with RMD: walking time per weekday, total travel time per weekday and regularity of doing physical leisure activity. In reference to walking less than 30 min per day, walking more than 30 min was significantly associated with higher RMD score with adjustments of other variables ($\boldsymbol{\beta} = 0.292$, 95%CI = 0.065–0.519 in model 1). In reference to travelling less than two hours per day, travelling more than two hours was significantly associated with lower RMD score adjusting for other variables ($\boldsymbol{\beta} = -0.282$, 95%CI = −0.506 to −0.058 in model 1). In reference to regular physical leisure activity, no regular physical leisure activity was significantly associated with lower RMD score after the adjustments ($\boldsymbol{\beta} = -0.276$, 95%CI = −0.533 to −0.018 in model 1) (Table 6.14). Thus, the statistical results suggested that walking more than 30 min daily, travelling less than two hours per day, and regular leisure activity were associated with lower risks of mental disorders.

Table 6.14 Results of linear regression models examining association between RMD and health-related residential behaviors

Model predictors	Model 1 β (95%CI) p (N = 292)	Model 2ª OR (95%CI) (N = 382)
Physical activity-travel activity and active leisure activity		
Travel mode for commuting		
Active travel mode—Reference		
Public transportation	−0.048 (−0.298, 0.202) p = 0.703	
Private car	0.245 (−0.063, 0.553) p = 0.119	
Multiple choice	0.138 (−0.237, 0.513) p = 0.469	
Walking per weekday		
<30 min—Reference		
≥30 min	0.292* (0.065, 0.519) p = 0.012	
Total travel time per weekday		
<2 h—Reference		
≥2 h	−0.282* (−0.506, −0.058) p = 0.014	
Willingness for walking: Compared to other means, I prefer to walk to destinations in daily life		
Disagree—Reference		
Not sure	0.159 (−0.195, 0.513) p = 0.377	
Agree	0.179 (−0.126, 0.483) p = 0.249	
Leisure activity		
Variety of leisure activities		
None—Reference		
One type	0.022 (−0.201, 0.244) p = 0.848	
Two types	0.033 (−0.239, 0.305) p = 0.810	
Preference for leisure activities: I love doing leisure activities such as physical exercises		
Agree—Reference		
Not sure	0.042 (−0.253, 0.336) p = 0.780	
Disagree	0.018 (−0.296, 0.332) p = 0.910	
Regularity of doing leisure activities: I did physical exercises regularly during past six months		
Agree—Reference		
Not sure	−0.127 (−0.377, 0.124) p = 0.321	
Disagree	−0.276* (−0.533, −0.018) p = 0.036	

(continued)

Table 6.14 (continued)

Model predictors	Model 1 β (95%CI) p (N = 292)	Model 2[a] OR (95%CI) (N = 382)
Neighbors' interaction		
Variety of informal neighbors' interactions		
1 type and less—Reference		
2 types		0.099 (−0.128, 0.325) p = 0.393
3 types and more		0.229* (0.011, 0.448) p = 0.040
Frequency of informal neighbors' interactions		
Every day—Reference		
More than 3 times per week		−0.127 (−0.350, 0.097) p = 0.266
Once to twice per week		−0.134 (−0.357, 0.088) p = 0.235
Once per month		−0.133 (−0.432, 0.166) p = 0.383
Never		−0.045 (−0.377, 0.286) p = 0.787
Willingness of joining informal neighbors' interactions: In my spare time, I'd rather to stay at home than to interact with my neighbors		
Agree—Reference		
Not sure		−0.011 (−0.276, 0.254) p = 0.936
Disagree		0.194 (−0.015, 0.402) p = 0.069
Variety of community activities		
None—Reference		
1 types		−0.041 (−0.297, 0.214) p = 0.751
2 types and more		−0.179 (−0.598, 0.240) p = 0.402
Frequency of attending community activities		
Never—Reference		
Once per month or less		−0.137 (−0.385, 0.112) p = 0.280
More than once per month		−0.036 (−0.323, 0.251) p = 0.807
Willingness of attending community activities: I have no interests in the community activities		
Agree—Reference		

(continued)

Table 6.14 (continued)

Model predictors	Model 1 β (95%CI) p (N = 292)	Model 2[a] OR (95%CI) (N = 382)
Not sure		0.074 (−0.142, 0.289) p = 0.501
Disagree		0.077 (−0.131, 0.285) p = 0.469
Individual socio-demographics		
Age		
20–29—Reference		
30–39	−0.009 (−0.259, 0.242) p = 0.946	−0.024 (−0.240, 0.192) p = 0.826
40–49	0.116 (−0.217, 0.448) p = 0.494	0.071 (−0.213, 0.356) p = 0.623
50–59	0.391 (−0.003, 0.785) p = 0.052	0.429* (0.083, 0.774) p = 0.015
≥60	0.016 (−0.433, 0.466) p = 0.943	0.078 (−0.309, 0.4664) p = 0.693
Sex		
Female	−0.137 (−0.325, 0.052) p = 0.155	−0.129 (−0.294, 0.036) p = 0.125
Income		
< 4w—Reference		
4−8w	0.087 (−0.181, 0.355) p = 0.524	0.004 (−0.234, 0.242) p = 0.974
8−12w	0.099 (−0.207, 0.404) p = 0.525	0.018 (−0.251, 0.287) p = 0.895
12w	−0.125 (−0.434, 0.183) p = 0.425	−0.107 (−0.377, 0.164) p = 0.438
Residence length		
≥5 year—Reference		
< 1 year	0.041 (−0.358, 0.440) p = 0.839	0.055 (−0.297, 0.406) p = 0.759
1–2 year	0.262 (−0.040, 0.564) p = 0.089	0.222 (−0.049, 0.493) p = 0.109
2–3 year	0.158 (−0.118, 0.4343) p = 0.261	0.138 (−0.104, 0.3979) p = 0.263
3–4 year	0.120 (−0.152, 0.392) p = 0.385	0.163 (−0.071, 0.397) p = 0.170
4–5 year	0.024 (−0.328, 0.377) p = 0.892	0.043 (−0.261, 0.346) p = 0.782
Education status		

(continued)

Table 6.14 (continued)

Model predictors	Model 1 β (95%CI) p (N = 292)	Model 2[a] OR (95%CI) (N = 382)
Bachelor or above—Reference		
Primary school and below	0.053 (−0.395, 0.502) p = 0.815	0.052 (−0.321, 0.426) p = 0.784
Secondary school	−0.092 (−0.457, 0.274) p = 0.622	−0.140 (−0.436, 0.155) p = 0.351
High school	−0.132 (−0.447, 0.183) p = 0.411	−0.102 (−0.356, 0.152) p = 0.430
Junior college	0.018 (−0.260, 0.295) p = 0.901	0.004 (−0.233, 0.242) p = 0.971
Employment status		
No jobs—Reference		
Have to work	−0.196 (−0.524, 0.131) p = 0.239	−0.135 (−0.427, 0.157) p = 0.363
Family structure		
One generation—Reference		
Two-generation	0.076 (−0.168, 0.320) p = 0.538	0.094 (−0.118, 0.306) p = 0.383
Three-generation	0.163 (−0.128, 0.455) p = 0.271	0.205 (−0.045, 0.455) p = 0.108
Others	−0.130 (−0.541, 0.282) p = 0.535	−0.145 (−0.502, 0.213) p = 0.427
Homeownership		
No	−0.150 (−0.375, 0.075) p = 0.190	−0.155 (−0.353, 0.042) p = 0.122
R^2	0.209	0.182
F	F (35, 257) = 1.938 p = 0.002	F (36, 346) = 2.138 p < 0.001

The associations between walking, regular physical leisure activity and RMD are related to the effects of physical activity on mental health. According to the review in Chap. 2, studies have consistently and universally found that physical activity could benefit emotional well-being and relieve symptoms of mental disorders [15, 16, 26, 39]. Ohta et al. [36] note that leisure-time exercise and active travelling to work may be associated with better mental health in Japanese men with controlling job stress. In addition, the intervention effects of physical activity have also been proved in literature: physical activity or exercise programs can improve mood and reduce symptoms of depression and anxiety [46]. People diagnosed with depression, aerobic-exercise has been proven a better intervention than psychotropic treatment [3].

In this study, since the measurement for the regularity of engaging in physical leisure activity was self-reported, it is vulnerable to reporting bias. This makes it hard to determine which is the cause or effect in the association between the regular physical leisure activity and the prevalence of depression. However, supported by the literature, it is inferred that both walking and regular physical leisure activity could reduce the risks of mental disorders. Interviews with residents also confirmed the associations.

> If I feel down, I like to walk around. Walking could relax me. (GV_FO4)

> When I feel bad, I usually go for a walk or dance with neighbors. Then I feel much better and have a good mood. This could also help me to sleep well at night. (AG_FY4)

> If I have the time, I walk rather than taking vehicles. Because I pay much attention to my health, I want to use every opportunity to do physical exercise. (LNT_FM8)

> Usually, I play Ping-Pong once to twice per week and I jog five days per week. Regular physical exercise makes me keep a good mental state. If my routine is interrupted, I feel bad emotionally since I am used to the regular physical activity. (LNT_MY6)

In addition, total travel time is determined by the distance and connectivity of destinations, such as from job to home, from home to hospital, shopping center, and schools, and the mode of travel. Previous studies support the relationship between travel time and RMD. Moist et al. [34] argue that longer travel time could decrease health-related quality of life. Literature on the suburban sprawl and mental health points out that the opportunity structure in the suburbs constrains the amount of daily physical activity by long commuting, which is the obvious mechanism between the sprawling environment and health [52]. In addition, more time spent driving alone in car could reduce the informal social interactions that contribute to social capital, and this is particularly true in the US (Sturm and Cohen [52]). Thus, it is quite possible that excessive long travel time could reduce opportunities for physical activity and interacting with neighbors, which undermine mental health. Interviews with residents reveal the relations between travel time and risks of mental disorders.

> The negative impact of excessive travel time is very obvious for me, which could increase the feeling of anxiety. Especially when I miss a bus and spend another 25 minutes waiting for another one. (AG_MY3)

> I have to spend around three hours commuting every day, which induces a high level of anxiety. Because it is not very comfortable on the bus or on the subway and you could do nothing. Sometimes I have to stand for a long time since there are so many passengers on line 3. (AG_FY8)

> The travel time could influence the quality of life indeed. After I change my work place from the one in Huadu district (which is very far from AG) to the nearby one, I have been much happier than before, since I do not need to spend so much time on commuting. It is really a waste of life spending too much time on the way, which could cause a bad mood. (AG_FM1)

These statements make two points. A two-hour commute reduced the amount of time that could be spent on physical leisure activity and interaction with neighbors. The trips themselves are tiresome and difficult, worsening travelers' mood and increasing their anxiety.

6.5.2.2 Associations Between Neighbors' Interaction and RMD

Only one factor derived from neighbors' interaction was found significantly associated with RMD. In reference to participating in one or less type of informal interaction with neighbors, participating in three or more types was significantly associated with higher RMD score with adjustment of other variables ($\beta = 0.229$, 95%CI = 0.011–0.448) (model 2 in Table 6.8). Thus, engaging in various informal interactions with neighbors was associated with low risks of mental disorders.

To explain this association, two mediators, social support and social cohesion, could be introduced, with the assumption that more types of neighbors' interactions may increase the opportunities for getting social support from the neighborhood and nurturing social cohesion, both of which benefit mental health. There was an average of 2.4 choices for each respondent in terms of the informal interactions. Several types of interactions were related to information exchange, such as chatting face-to-face and online (40.9%), babysitting (15.1%), playing games (8.5%), doing physical exercise (12.4%), and helping others (19.9%) (Fig. 6.16). More types of informal interactions could enable people to obtain information from others, make new friends, grow a social network and obtain social support.

As discussed in Chap. 2, many studies have shown the importance of social support for physiological health [51]. In the US, less socially cohesive neighborhoods were associated with increased depression, smoking, and not walking for exercise [12]. In the Chinese context, Cao et al. [8] find that trust, reciprocity, and social network were significantly associated with the geriatric depression while social participation

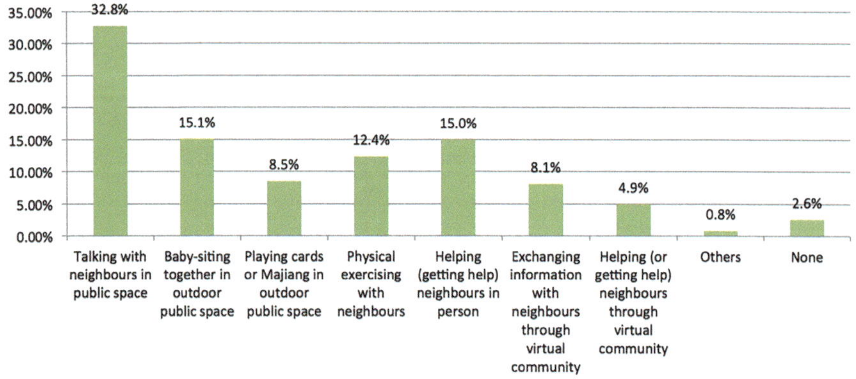

Fig. 6.16 Types of informal interaction with neighbors (N = 1278)

was not correlated. Although the findings are not consistent in different cultural contexts, studies have affirmed that good social relations and social capital contribute to good emotional wellbeing [44, 48, 58]. Thus, this study proposes that participating in various informal interactions may reduce risks of mental disorders by gaining social support, enjoying good neighborhood cohesiveness. In interviews, residents suggested that informal interactions could resolve emotional problems and cultivate good relationships with neighbors. As four residents remarked:

> I like to come to the public space and watch neighbors to play cards, and we chat. If I could not find anybody to chat with, I would feel very annoyed. Interacting with neighbors always make me feel better. If I could talk to my neighbors, I feel very happy. (GV_FO4)

> Communicating with neighbors could help to relieve stress. Sometimes I get myself into a dead end due to something and believe it is really a serious issue, however, when you talk it to others, you could get some advice and learn from others' experience. Then the big issue could be completely dispelled as an issue. (LNT_FY7)

> I have intentions to communicate with people. When I feel depressed, I'd like to talk to others. Talking about the bothering issues with others could help me to shift my attention or even inspire me. I'd like to talk to people, including neighbors, when I feel down. (LNT_FM8)

> I like to interact with neighbors every day. Since I do not have many relatives and friends here, I desire to get to know neighbors, which could help me feel less stressful. (LNT_FY1)

These comments reveal that the benefits of the informal interactions lie in obtaining suggestions, inspiration comfort and moral support. In most cases, neighbors could compensate for absent friends or family members in terms of resolving psychological problems. Thus, the associations that engaging in diverse informal interactions could contribute to lower risks of mental disorders is largely confirmed, highlighting the importance of informal interaction with neighbors for keeping a good mental state.

6.6 Conclusion

This chapter explores the physical and mental health in association with the suburban living. Factors formulated by the planning and design of this specific residential form in suburbs and generated by the resultant suburban lifestyle of the middle-income population are scrutinized.

First, factors influencing physical health are summarized. For the environmental determinants, this chapter finds that the differences of physical characteristics across the three SCHEs have no effect on self-reported general physical health, and these differences appear in the pedestrian-vehicles relation, street density and morphology, public transportation system, quantity and distribution of green space, quality of schools and daily commercial services, the sense of belonging and aesthetic style. It is argued that these differences may have similar effects on the self reported general physical health, or they have no impact at all, indicating the necessity of further study on the effects of residential planning and design on residents' physical health. For the behavioral determinants, walking more than 30 min per weekday was associated with the lower prevalence of obesity; driving more than one hour per weekday and preferring doing physical leisure activity were significantly associated with higher prevalence of feeling in good general physical health. Based on this study, although it is debatable whether walking is the reason of reducing the prevalence of obesity, it is ascertained that physical activity is inherently beneficial for both preventing obesity and promoting general physical health.

Second, for the factors influencing depression, AG was significantly associated with higher odds of being depressed, primarily due to the selection mechanism. Spending more than 30 min on public transportation was significantly associated with higher odds of being depressed, which is mainly explained by the psychosocial stress from poor travel conditions. Participating in at least one type physical leisure activity was associated with much lower odds of being depressed, showing the benefits of physical activity. For risks of mental disorders, influential factors were identified in Table 6.15. Notably, living in AG and in a villa or townhouse was associated with lower risks of mental disorders than living in LNT and in high-rise buildings (17 floors), respectively. This is explained by the reflection of a home on the quality of living environment and the socioeconomic status, which in turn may affect the sense of identity, self-esteem and control over one's life. Being within walking distance of destinations was associated with lower risks of mental disorders, which is determined by the health effects of walking. Convenient and attractive places for physical leisure activity within the neighborhood could increase the possibility of engaging in physical activity, reducing risks of mental disorders. Diverse and available educational facilities and daily commercial facilities could lower the risks of mental disorders through releasing the social stressors. Green space was associated with stress release, physical activity and neighbors' interaction, decreasing risks of mental disorders. Having a sense of belonging could reduce loneness and generating neighborhood cohesion, lowering the risks of mental disorders. Walking more than 30 min, traveling less than two hours, and doing regular physical leisure activity were associated with low risks of mental disorders by increasing the amount of physical activity and opportunities for interactions with neighbors. Participating in more informal interactions may reduce the risks of mental disorders by gaining social support and enjoying good neighborhood cohesion (Table 6.15).

On the whole, four groups of associations are visualized in Fig. 6.17, with the one color indicating one group of associations (red for obesity, orange for self-reported general physical health, blue for depression, and green for RMD). The lines refer to the connections between the items in both ends with the influential directions, and the width of the lines indicate the strength of the effects when treat each influential factor equally in terms of weight. In particular, six explanations for the factors affecting physical and mental health are identified (Fig. 6.17). Except the selection mechanism, there are five mediators within the causal mechanisms: daily routine, benefits of physical activity, psychosocial stress and life stressor, identity and symbolic meaning and self-esteem, and social support and neighborhood cohesion (Fig. 6.17).

Table 6.15 Environmental and behavior factors influencing RMD

Domains	Factors of risks of mental disorders
Physical residential environment	• Neighborhood type • Housing type
Walking environment	• Accessibility of destinations within neighborhood
Leisure environment	• Place preference for leisure activity
Service environment	• Satisfaction with educational facilities • Satisfaction with daily commercial facilities
Intangible environment	• Quantity of green space
	• Sense of belonging
Physical activity	• Walking time per day • Overall travel time per day • Self-report regularity of doing leisure activity
Neighbors' interaction	• Variety of informal neighbors' interaction

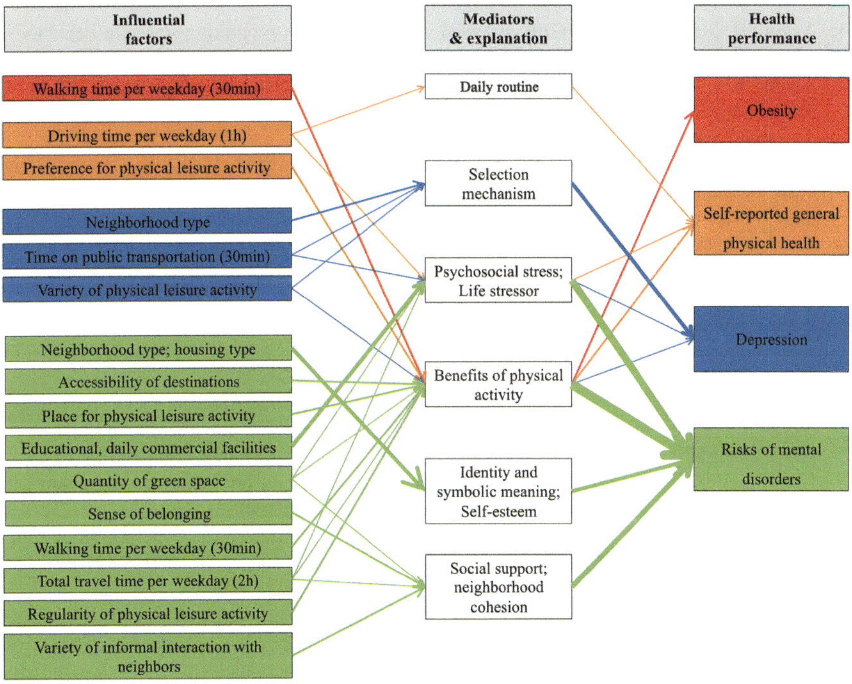

Fig. 6.17 Explanation between influential factors and health performance

Therefore, this chapter ascertains the health effects of physical activity on physical health, and health implications of the residential environment of SCHEs, physical activity, and neighbors' interactions on mental health. Putting up with the limitation of this study, direct effects of residential environment on the risks of mental disorders are affirmed, and indirect impact of residential environment through affecting residents' behaviors could be inferred, which will be explained in the next chapter in relation to residents-environment transactions. For the resultant way of life, investigations on travel and leisure activities indicate that the way of life in SCHEs could not be regarded as either sedentary lifestyle or health-promoting lifestyle. Thus, it is pointed out that several distinctive attributes of suburban living have crucial influences on residents' health, such as the negative impacts of the long-distance commuting and unsatisfied public service facilities. Compared with living environment in urban core areas, the suburban neighborhoods have more advantages in terms of the opportunities for physical leisure activity, greenery space and aesthetic landscape, and the sense of belonging, which contribute to the residents' health.

References

1. Acton GJ, Malathum P (2000) Basic need status and health-promoting self-care behavior in adults. West J Nurs Res 22(7):796–811. https://doi.org/10.1177/01939450022044764
2. Adams KB, Sanders S, Auth EA (2004) Loneliness and depression in independent living retirement communities: risk and resilience factors. Aging Ment Health 8(6):475–485. https://doi.org/10.1080/13607860410001725054
3. Babyak M, Blumenthal JA, Herman S, Khatri P, Doraiswamy M, Moore K, Edward Craighead W, Baldewicz TT, Ranga Krishnan K (2000) Exercise treatment for major depression: maintenance of therapeutic benefit at 10 months. Psychosom Med 62(5):633
4. Bauman A, Allman-Farinelli M, Huxley R, James WPT (2008) Leisure-time physical activity alone may not be a sufficient public health approach to prevent obesity—a focus on China. Obes Rev 9(s1):119–126. https://doi.org/10.1111/j.1467-789X.2007.00452.x
5. Baumeister RF, Leary MR (1995) The need to belong: desire for interpersonal attachments as a fundamental human motivation. Psychol Bull 117:497–529. https://doi.org/10.1037/0033-2909.117.3.497
6. Björk J, Albin M, Grahn P, Jacobsson H, Ardö J, Wadbro J, Östergren PO, Skärbäck E (2008) Recreational values of the natural environment in relation to neighbourhood satisfaction, physical activity, obesity and wellbeing. J Epidemiol Community Health 62(4):e2–e2
7. Booth KM, Pinkston MM, Poston WSC (2005) Obesity and the built environment. J Am Diet Assoc 105(5):110–117. https://doi.org/10.1016/j.jada.2005.02.045
8. Cao W, Li L, Zhou X, Zhou C (2015) Social capital and depression: evidence from urban elderly in China. Aging Ment Health 19(5):418–429. https://doi.org/10.1080/13607863.2014.948805
9. Cohen J (1992) A power primer. Psychol Bull 112:155–159. https://doi.org/10.1037/0033-2909.112.1.155
10. Cummings SM, Cockerham C (2004) Depression and life satisfaction in assisted living residents. Clin Gerontol 27(1–2):25–42. https://doi.org/10.1300/J018v27n01_04
11. De Vries S, Verheij RA, Groenewegen PP, Spreeuwenberg P (2003) Natural environments—healthy environments? An exploratory analysis of the relationship between greenspace and health. Environ Plan A 35(10):1717–1731
12. Echeverría S, Diez-Roux AV, Shea S, Borrell LN, Jackson S (2008) Associations of neighborhood problems and neighborhood social cohesion with mental health and health behaviors: the multi-ethnic study of atherosclerosis. Health Place 14(4):853–865. https://doi.org/10.1016/j.healthplace.2008.01.004

13. Ekers M, Hamel P, Keil R (2012) Governing suburbia: modalities and mechanisms of suburban governance. Reg Stud 46(3):405–422. https://doi.org/10.1080/00343404.2012.658036
14. Evans GW (2003) The built environment and mental health. J Urban Health 80(4):536–555. https://doi.org/10.1093/jurban/jtg063
15. Frank L, Engelke PO (2001) The built environment and human activity patterns: exploring the impacts of urban form on public health. J Plan Lit 16(2):202–218. https://doi.org/10.1177/08854120122093339
16. Frank L, Engelke P, Schmid T (2003) Health and community design: the impact of the built environment on physical activity. Island Press
17. Frank L, Kerr J, Sallis JF, Miles R, Chapman J (2008) A hierarchy of sociodemographic and environmental correlates of walking and obesity. Prev Med 47(2):172–178. https://doi.org/10.1016/j.ypmed.2008.04.004
18. Frumkin H (2002) Urban sprawl and public health. Public Health Rep 117
19. Groenewegen PP, van den Berg AE, Maas J, Verheij RA, de Vries S (2012) Is a green residential environment better for health? If so, why? Ann Assoc Am Geogr 102(5):996–1003. https://doi.org/10.1080/00045608.2012.674899
20. Hagerty BM, Williams A (1999) The effects of sense of belonging, social support, conflict, and loneliness on depression. Nurs Res 48(4):215
21. Halpern D (1995) Mental health and the built environment: more than bricks and mortar? Taylor & Francis US
22. Haslam SA, Jetten J, Postmes T, Haslam C (2009) Social identity, health and well-being: an emerging agenda for applied psychology. Appl Psychol Int Rev Psychol Appl Revue Int 58(1):1–23. https://doi.org/10.1111/j.1464-0597.2008.00379.x
23. Hill DL (2006) Sense of belonging as connectedness, American Indian worldview, and mental health. Arch Psychiatr Nurs 20(5):210–216. https://doi.org/10.1016/j.apnu.2006.04.003
24. Hu G, Pekkarinen H, Hänninen O, Yu Z, Guo Z, Tian H (2002) Commuting, leisure-time physical activity, and cardiovascular risk factors in China. Med Sci Sports Exerc 34(2):234
25. Jackson LE (2003) The relationship of urban design to human health and condition. Landsc Urban Plan 64(4):191–200. https://doi.org/10.1016/S0169-2046(02)00230-X
26. Jackson RJ, Kochtitzky C (2003) Creating a healthy environment: the impact of the built environment on public health. In: Sprawl watch clearinghouse monograph series, vol 19
27. Kawachi I, Berkman LF (2001) Social ties and mental health. J Urban Health 78(3):458–467. https://doi.org/10.1093/jurban/78.3.458
28. Levine JA, McCrady SK, Lanningham-Foster LM, Kane PH, Foster RC, Manohar CU (2008) The role of free-living daily walking in human weight gain and obesity. Diabetes 57(3):548–554. https://doi.org/10.2337/db07-0815
29. Lin N, Lai G (1995) Urban stress in China. Soc Sci Med 41(8):1131–1145. https://doi.org/10.1016/0277-9536(94)00426-T
30. Maas J, Verheij RA, Groenewegen PP, de Vries S, Spreeuwenberg P (2006) Green space, urbanity, and health: how strong is the relation? J Epidemiol Community Health 60(7):587–592. https://doi.org/10.1136/jech.2005.043125
31. Maslow AH, Frager R, Fadiman J, McReynolds C, Cox R (1970) Motivation and personality, vol 2. Harper & Row
32. Miles R, Panton LB, Jang M, Haymes EM (2008) Residential context, walking and obesity: two African-American neighborhoods compared. Health Place 14(2):275–286. https://doi.org/10.1016/j.healthplace.2007.07.002
33. Mitchell R, Popham F (2007) Greenspace, urbanity and health: relationships in England. J Epidemiol Community Health 61(8):681–683
34. Moist LM, Bragg-Gresham JL, Pisoni RL, Saran R, Akiba T, Jacobson SH, Fukuhara S, Mapes DL, Rayner HC, Saito A, Port FK (2008) Travel time to dialysis as a predictor of health-related quality of life, adherence, and mortality: the dialysis outcomes and practice patterns study (DOPPS). Am J Kidney Dis 51(4):641–650. https://doi.org/10.1053/j.ajkd.2007.12.021
35. Nielsen TS, Hansen KB (2007) Do green areas affect health? Results from a Danish survey on the use of green areas and health indicators. Health Place 13(4):839–850

36. Ohta M, Mizoue T, Mishima N, Ikeda M (2007) Effect of the physical activities in leisure time and commuting to work on mental health. J Occup Health 49(1):46–52. https://doi.org/10.1539/joh.49.46
37. Paeratakul S, White MA, Williamson DA, Ryan DH, Bray GA (2002) Sex, race/ethnicity, socioeconomic status, and BMI in relation to self-perception of overweight. Obes Res 10(5):345–350. https://doi.org/10.1038/oby.2002.48
38. Papas MA, Alberg AJ, Ewing R, Helzlsouer KJ, Gary TL, Klassen AC (2007) The built environment and obesity. Epidemiol Rev 29(1):129–143. https://doi.org/10.1093/epirev/mxm009
39. Penedo FJ, Dahn JR (2005) Exercise and well-being: a review of mental and physical health benefits associated with physical activity. Curr Opin Psychiatry 18(2):189
40. Peng C-YJ, Lee KL, Ingersoll GM (2002) An introduction to logistic regression analysis and reporting. J Educ Res 96(1):3–14. https://doi.org/10.1080/00220670209598786
41. Popkin BM (2008) Will China's nutrition transition overwhelm its health care system and slow economic growth? Health Aff 27(4):1064–1076. https://doi.org/10.1377/hlthaff.27.4.1064
42. Pretty J, Peacock J, Sellens M, Griffin M (2005) The mental and physical health outcomes of green exercise. Int J Environ Health Res 15(5):319–337. https://doi.org/10.1080/09603120500155963
43. Prieto-Flores M-E, Fernandez-Mayoralas G, Forjaz MJ, Rojo-Perez F, Martinez-Martin P (2011) Residential satisfaction, sense of belonging and loneliness among older adults living in the community and in care facilities. Health Place 17(6):1183–1190. https://doi.org/10.1016/j.healthplace.2011.08.012
44. Renalds A, Smith TH, Hale PJ (2010) A systematic review of built environment and health. Fam Community Health 33(1):68–78. https://doi.org/10.1097/FCH.0b013e3181c4e2e5
45. Rosenbaum JE, Reynolds L, Deluca S (2002) How do places matter? The geography of opportunity, self-efficacy and a look inside the black box of residential mobility. Hous Stud 17(1):71–82. https://doi.org/10.1080/02673030120105901
46. Ross CE, Hayes D (1988) Exercise and psychologic well-being in the community1. Am J Epidemiol 127(4):762–771. https://doi.org/10.1093/oxfordjournals.aje.a114857
47. Sargent J, Williams RA, Hagerty B, Lynch-Sauer J, Hoyle K (2002) Sense of belonging as a buffer against depressive symptoms. J Am Psychiatr Nurses Assoc 8(4):120–129. https://doi.org/10.1067/mpn.2002.127290
48. Sarkar C, Webster C, Gallacher J (2014) Healthy cities: public health through urban planning. Edward Elgar Publishing
49. Shen Y-C, Zhang M-Y, Huang Y-Q, He Y-L, Liu Z-R, Cheng H, Tsang A, Lee S, Kessler RC (2006) Twelve-month prevalence, severity, and unmet need for treatment of mental disorders in metropolitan China. Psychol Med 36(2):257–267. https://doi.org/10.1017/S0033291705006367
50. Stephens T (1988) Physical activity and mental health in the United States and Canada: evidence from four population surveys. Prev Med 17(1):35–47. https://doi.org/10.1016/0091-7435(88)90070-9
51. Stokols D (1992) Establishing and maintaining healthy environments: toward a social ecology of health promotion. Am Psychol 47:6–22. https://doi.org/10.1037/0003-066X.47.1.6
52. Sturm R, Cohen DA (2004) Suburban sprawl and physical and mental health. Public Health 118(7):488–496. https://doi.org/10.1016/j.puhe.2004.02.007
53. Sullivan WC, Chang C-Y (2011) Mental health and the built environment. In: Dannenberg AL, Frumkin H, Jackson RJ (eds) Making healthy places: designing and building for health, well-being, and sustainability. Island Press/Center for Resource Economics, pp 106–116. https://doi.org/10.5822/978-1-61091-036-1_7
54. Thoits PA (1985) Social support and psychological well-being: theoretical possibilities. In: Sarason IG, Sarason BR (eds) Social support: theory, research and applications. Springer Netherlands, pp 51–72. https://doi.org/10.1007/978-94-009-5115-0_4
55. Verheij RA (1996) Explaining urban-rural variations in health: a review of interactions between individual and environment. Soc Sci Med 42(6):923–935. https://doi.org/10.1016/0277-9536(95)00190-5

56. Walker SN, Sechrist KR, Pender NJ (1987) The health-promoting lifestyle profile: development and psychometric characteristics. Nurs Res 36(2):76–81
57. Weich S, Blanchard M, Prince M, Burton E, Erens B, Sproston K (2002) Mental health and the built environment: cross-sectional survey of individual and contextual risk factors for depression. Br J Psychiatry 180(5):428–433. https://doi.org/10.1192/bjp.180.5.428
58. Wilkinson RG, Marmot M (2003) Social Determinants of health: the solid facts. World Health Organization

Chapter 7
Explanation of Factors Influencing Health: Residents-Environment Transactions and Institutional Factors

Factors associated with physical and mental health have been identified from residential environment and residential behavior, respectively. However, the mutual effects of environmental and behavioral factors have not been considered in previous analysis chapters. As discussed in Chap. 2, people-environment congruity is the essence of a health-promoting environment, so examination of residents-environment transactions could help to understand the interactions between environmental and behavioral factors and to identify the issues in the creation of a health-promoting environment. Furthermore, since the residential environmental factors that influence health are determined by housing development, it is argued that institutional factors in housing development could have great impact on the formation of health-promoting environments. Thus, this chapter aims to explain how the factors affecting health have been formulated through people-environment transactions in suburban living and how the residential environment has been affected by the SCHEs development institutionally.

This chapter is comprised of two analytical sections. The first section focuses on the transactions between residents and their environment from the human ecological perspective. The interpretive, responsive, and operative transaction between residents of SCHEs and their environment are examined. It is noted that suburban residents passively adapt to their residential environment. In other words, the residential environment determines health-related behaviors and residents are powerless to change it. Thus, the passive transaction between residents and the suburban living environment could threaten their health finally. The second part centers on the institutional causes of the threats to health during suburban housing development. The roles of the state, the market and the planners and their interactions are examined in suburban housing development, focusing on the stages of developing initiatives and decision-making. Finally, institutional reasons that affect the factors influencing health are discussed, focusing on the provision of public service and residents' participation in housing management.

T. Zhang, *Towards Healthy Settlements*, Urban Sustainability,
https://doi.org/10.1007/978-981-97-1207-6_7

7.1 Residents-Environment Transactions

This section therefore examines the residents-environment transactions in suburban living, focusing on the factors influencing health that are identified in Chaps. 5 and 6. Three transaction modes are examined and explained: interpretive (physical residential environment and perceived residential environment), responsive (perceived residential environment and health-related residential behavior), and operative (health-related residential behavior and physical residential environment). Both qualitative analysis and quantitative analysis are used to comprehend the transaction processes.

7.1.1 Interpretive Transaction: Cognition on Residential Environment

Interpretive people-environment transaction focuses on residents cognitively constructing and interpreting their environment [14]. This section explains how residents cognize and perceive the physical residential environment and formulate perceived residential environments, which significantly influence the risks of mental disorders in suburban living.

The concept of image cognition of the places is borrowed to explain the interpretive transactions between the physical surroundings and residents' cognition and perception. Images are "representations of reality of space, time, relations, persons, roles and organizations" [1, p. 273]. Human beings use their cognitive representations of various physical settings, the images, to organize information, to guide their behaviors, and to extract or store real-world scenes, forming beliefs and knowledge [1, 3]. In addition, the images evolve according to the way we value and use the environment, so our actions in the environment could modify our images. Thus, it is argued that the interaction between the images and behaviors is the essence of the cognitive process [1].

In addition, the cognition and perception of the same physical setting may vary among populations. A variety of personal and cultural factors are important in construing the environment and expressing perceptions, such as gender, socioeconomic status and ethnic identity [1, 8, 14]. The cognitive images of a given place vary across observers due to individual's uniqueness in experience [8]. However, shared physical reality, physiology, and culture could generate areas of agreement in cognitive images [8]. Therefore, demographic attributes, socioeconomic status, individual's experiences and demands have crucial influence on the image cognition of places, which may lead to different subgroups of residents interpreting the same physical residential environment in different ways.

Six factors of perceived residential environment are identified in Chap. 6: accessibility of destinations, place preference for physical leisure activity, educational and daily commercial facilities, green space, and sense of belonging. These factors also appear at the studies on the image cognition of residential environment, such as what

Banerjee and Baer [1] has argued: public facilities such as schools, parks and playgrounds are significant components of the images of the residential environment, so do the commercial facilities such as super markets, drug stores, and quick-order restaurants. Thus, it makes sense to explain the interpretive transaction with the help of image cognition. The following passages focus on the image cognition of residents on public service facilities and design features, which affect the risks of mental disorders significantly.

7.1.1.1 Public Service Facilities

As shown in Chap. 6, both educational facilities and daily commercial facilities are significantly associated with the risks of mental disorders, which residents described as social stressors. The extent to which the quantity and quality of the service facilities could meet the needs of residents affects their cognition and perception of the immediate living environment, affecting their perceived stress.

For educational facilities, Fig. 6.2 suggests that there is one public primary school and one public middle school around LNT. Figure 6.3 shows that there are two public primary schools, one public middle school and one private middle school around AG. Figure 6.4 shows that there is one public primary school and one public middle school in GV. Section 6.4.2 revealed AG residents' strong dissatisfaction with their schools, highlighting residents' concerns on the quality of schools in suburbs of Guangzhou. Based on previous analyses, this section compares cognitions of residents in the three SCHEs on schools.

Such opinions show that schools for LNT and AG residents are dissatisfactory, while those for GV residents are an attraction for residents. Interviews suggested that two greatest cognition and worries among LNT and AG residents are available places in schools and schools' quality. The government regulation of schools and questions about educational quality exacerbate the dissatisfaction of residents. Some studies suggest that residents tend to trust the quality of public schools more than they do the private schools built by developers in suburban areas, so many residents choose to send their children to high-quality public schools in the urban core area [7]. In addition, private schools are more expensive to attend, which is another reason for choosing public schools [6, 7]. Therefore, for AG residents, although there are three public schools near AG, only one is available for them. Even for the AG residents who could afford the private primary schools, they were worried about the educational quality. To the contrary, GV residents are less concerned since the children of homeowners in GV are eligible for the public primary and middle schools and the educational quality is guaranteed.

Furthermore, comments in Table 7.1 suggested that the socio-demographic attributes such as age and homeownership are also determinative of cognition. Obviously, parents of school-age children were more concerned about school choice, while tenants were constrained in the choices since the school places are associated with children's registered residence location that is usually bounded with homeownership. Thus, government regulation of schools, sociocultural understanding of the quality of education, affordability of schooling, and individual socio-demographic attributes could all affect the cognition of residents on the schools near their neighborhoods.

In terms of the cognition on daily commercial facilities, the mismatches between residents' demands and market supplies could be identified through observation and interviews. Figures 6.2, 6.3, 6.4 and 7.1 suggest that across the three SCHEs, LNT had the most commercial space in the form of shops along the streets, while GV had the least commercial space with only a few stores, which had to be supplemented by informal retails provided by villagers from near villages. For AG, the commercial service concentrated on the center of the neighborhood, which was inconvenient for residents living in the edges of neighborhood.

For the cognition, more AG residents (58.5%) were satisfied with the daily commercial facilities than LNT residents (49.1%) and GV residents (15.4%). Comments listed in Table 7.2 could give some explanations on both consistency and inconsistency between residents' cognition and visitors' observation. Although there were limited choices in AG with several markets and shops along the main street of the neighborhood, the high satisfaction of AG residents was depended on the nearby Wanda shopping mall, fulfilling most residents' shopping demands. In addition, LNT residents, especially the affluent ones, were not satisfied with the type of commercial activities provided in the neighborhood, leading to that only half of respondents reported satisfaction with plenty of shops (Table 7.2). Most GV residents were dissatisfied with commercial facilities due to the limited choices of shops, which is consistent with the author's observation.

In addition, Table 7.2 suggests that socioeconomic attributes affected residents demands, and individual's experience influenced the cognition on the same commercial facilities. It is proposed that age, income, and migration status influence the way of valuing and using commercial facilities (Table 7.2). The foci of the young and the elderly are different: the young wanted various choices of recreational facilities and the elderly were more concerned with affordability. In addition, migrants in LNT were more likely to be satisfied with daily commercial facilities than the homeowners in LNT (Case 1 compared with case 2 and 3 in Table 7.2). Migrants in GV, especially the elderly ones move to SCHEs for their children, tended to compare the commercial services of the neighborhood with those of their hometowns (Case 8 in Table 7.2). Thus, socioeconomic status and individual's experience affect residents' cognition on daily commercial facilities.

7.1.1.2 Design Features

Two factors of design features affecting risks of mental disorders are the foci of this section: green space and the sense of belonging. Referring to Table 6.3, GV had the highest green ratio of 54.2%, followed by the AG (36.8%), and LNT had the lowest green ratio of 20%. Figures 6.2, 6.3 and 6.4 suggest that there were two types of green space: public green space without access control and gated green space with access control. Different distributions of green space could be identified comparing the three figures: public green space in AG and GV is located at the center of the

Table 7.1 Statements on educational facilities in the three cases

Statements	Cognition	Neighborhood	Individual characteristic
1. I think the schools in Luoxi Island are good and I was studying in the middles schools here. But the demands are increasing and I feel that the middle school has been enlarged in recent years. (LNT-FY7)	Satisfactory	LNT	Young age, without child, homeowner
2. I don't think the quality of the schools here is good. I may move in future when my child is about to go to school. (LNT-FM5)	Unsatisfactory, unbearable	LNT	Middle aged, has a child, tenant
3. The quality of the schools here is not satisfactory to me. But we are migrants, it is difficult for us to choose the top schools, and we have no choice. (LNT-FM4)	Unsatisfactory, helpless	LNT	Middle aged, has a child, tenant
4. The quality of the private schools here is not very good. In addition, there is no choice of the public schools because there is only one. (AG-MY3)	Unsatisfactory	AG	Young age, without a child, homeowner
5. We need better and more affordable public schools. Actually, nowadays, most children have to go private school and some residents choose to move away from this neighborhood because of the lack of proper school. (AG-FM1)	Unsatisfactory, unbearable and worried	AG	Middle aged, has a child, homeowner
6. The educational resources are very limited, especially in terms of the kindergartens and primary schools. I worried about it very much as a young father	Unsatisfactory, worried and anxious	AG	Middle aged, have a child, homeowner
7. One of the reasons that we bought the property here is the school. All children of homeowners could go to the primary school here, and this standard is much easier for us to achieve than other situations since we are migrants. (GV-FM2)	Satisfactory	GV	Middle aged, have a child, homeowner
8. The schools are very near and convenient, both primary school and middle school. The quality of the schools is satisfactory as well. It is a major reason that we chose to live here. (GV-MM3)	Satisfactory	GV	Middle aged, have a child, homeowner

Fig. 7.1 Commercial facilities in the three SCHEs

neighborhoods, while that in LNT is located at the periphery; additionally, AG has a well-designed landscape avenue linking its main entrance and almost all residential clusters. Gated green space is even-distributed in the residential clusters of AG and GV, while few residential clusters in LNT have gated green space. Appendix 1 gives the images of green space in the three cases, showing that the quality of green space in LNT is the worst. In addition, Fig. 7.2 shows that more AG residents (69.3%) reported adequate green space than GV residents (55.3%) and LNT (50.3%). Thus, in general, residents' cognition on the quantity of green space was consistent with the observation. Residents' comments in Table 7.3 show their perception of green space, ascertaining consistency.

For the sense of belonging, image cognition plays a crucial role between emotional responses to the physical characteristics of residential environment. Psychologist points out that human emotional experiences of the physical environment could be quantified by the affective appraisal, such as individuals' judgments that they like a particular place [8, as cited in Russell and Snodgrass 1989]. This is an attribution to the built environment, which could be determined by design. Thus, this study argues that the sense of belonging is the results of residents' affective appraisal of their physical residential environment. According to interviews, Table 7.4 identifies the factors affecting the sense of belonging.

These opinions indicate that the sense of belonging was affected by safety, physical boundary, service facilities, aesthetic, and symbolic and social meaning of design. Thus, physical design is essential to create the sense of belonging in SCHEs. This argument could be explained by the predisposition of the human favoring coherence. Coherence could help human make sense of the surroundings for fostering survival and insuring safety. Good coherence is featured by "naturalness, good upkeep, open views, order and historic elements", indicating high social status or upper class [8, p. 134]. The likable features of good coherence could make identifiable and sensuous places, contributing to favorable emotional meaning for people [8]. Thus, this study infers that physical and symbolic design, especially the façade, gate, facilities and landscape, may affect residents' cognition on the coherence of their residential environment, resulting in emotional appraisals on safety, aesthetics and social meaning, which influence the sense of belonging.

To summarize, for interpretive transactions, socioeconomic characteristics such as age, income, homeownership, and hukou status (household registration status)

Table 7.2 Statements on daily commercial facilities in the three SCHEs

Statements	SPHE	Individual characteristic
1. It is very convenient to live here, especially for us housewives. I can just walk downstairs and buy goods. (LNT-FM5)	LNT	Middle aged, migrant
2. The turnover rate is very high, and you could always see that several stores that are decorated in anyone streets. The brand store could not survive here because most consumers are the low to middle income population. There are a lot of tenants, comprising the major consumers. (LNT-MM3)	LNT	Middle aged, with Guangzhou Hukou
3. It is a bit of chaotic because of the increasing business opportunities. With the increasing migrants, shops and vendors are increasing. But the management is not effective, leading to the lack of order. (LNT-FY7)	LNT	Young, with Guangzhou Hukou
4. The recreational facilities are not enough, so it is a little boring. It is lucky that the Wanda Plaza has opened, making up my demands. (AG-FY4)	AG	Young, with Guangzhou Hukou
5. I usually go to the nearby village to buy vegetables although there are supermarkets in the neighborhood. The price of the goods is much higher than that the traditional market. (AG-MO5)	AG	The elderly, migrant
6. The market is too small, and it could not satisfy the demands of the residents. Many residents go to the nearby town (Shilou town) to buy things. (GV-FO1)	GV	The elderly, migrant
7. It is very inconvenient of living here, especially in terms of the commercial facilities. If I want to have my hair cut or want some late-night munchies, I have to drive outside. (GV-MM3)	GV	Young, with Guangzhou Hukou
8. Comparing with my hometown, it is not convenient for shopping. I couldn't fully understand the standard mandarin, and the service staff could not understand me because of my accent. (GV-FO4)	GV	The elderly, migrant

lead to various interpretations on the same residential environment. So subgroup residents, such as the elderly, the children, and housewife, the tenants and migrants, who with similar socio-demographic characteristics and life experiences, would share

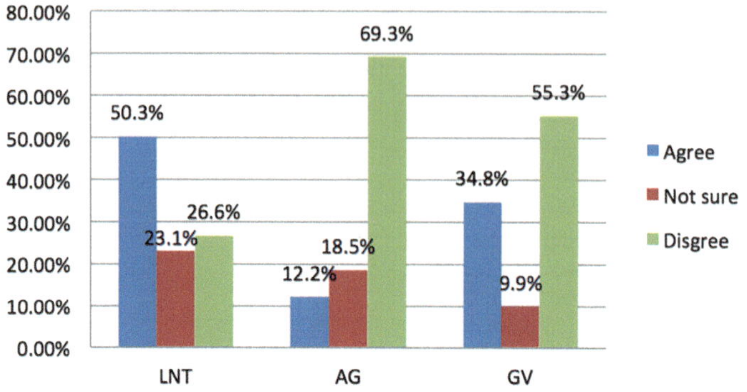

Fig. 7.2 Perception on green space: the green space is not enough in the neighborhood ($N = 515$)

Table 7.3 Statements on green space in the three SCHEs

Statements
1. Although there are few gardens, there are some parks, such as the Luoxi Park, which is free to the public. But we needs more green space. (LNT_FY7)
2. It will be good if there is another park because there is too less green space. (LNT_FM5)
3. The green space is enough, but the maintenance is becoming poorer than before. (AG_FO5)
4. We were attracted by the green space and landscape venue when we first visited this neighborhood. Then we decided to live here. (AG_MY3)
5. The green and open space is enough, which is quite good. (GV_FM2)

Table 7.4 Statements on the sense of belonging

Statement	Influential factors
1 It is definitely that safety could increase the sense of belonging of living here. (LNT-MM3)	Safety
2. This neighborhood is too open to generate the sense of belonging. If we had a physical boundary, the sense of belonging might be enhanced. Without a physical boundary, the sense of belonging may refer to the sense of the whole island (the Luoxi Island). (LNT- FY7)	Design of gate
3. There is no symbolic element in this neighborhood, making hard to have the sense of belonging. If there were some fancy and grand landmarks, it would be very proud of me to tell others that I live here. But we don' t have any, so you know, it is not a high-end neighborhood In addition, if the appearance of the buildings in my neighborhood could be unified and improved, it would be comfortable for looking. (LNT-FM5)	Symbolic and social meaning, aesthetic
4. The service facilities could make me feel the sense of belonging, especially the service of shuttle bus. Many residents living here take the shuttle bus to work and we have more chance to meet each other, increasing the neighbors' attachment. (AG-MY3)	Service facilities

similar perceptions and cognitions [8]. In addition, sociocultural understanding, individual experience and demands have crucial effects on residents' cognition. Moreover, coherence of the physical residential environment could affect the residents' sense of belonging through physical design. In particular, the visual characteristics mediate the interpretative transaction between residents' affective appraisal and physical residential environment. In this sense, interpretive transactions between residents and physical environment could influence residents' health through the match degree between cognition and expectation, indicating the importance of residents' demand for the physical residential environment in the creation of health-promoting environment.

7.1.2 *Responsive Transaction: Residential Environment and Physical Activity*

Responsive transaction highlights the effects of the environment on the individual's behavior [14]. This section focuses on how residents responded to their perceived residential environment in the form of health-related residential behaviors. Referring to previous findings, physical activity is influential for both physical and mental health. So the effects of the perceived residential environment on travel activity and physical leisure activity are examined.

7.1.2.1 Travel Activity: Travel Time of Different Modes

It has been found that time spent on walking, driving, public transportation and total travel time were associated with residents' health significantly. Both distance and travel mode influence travel time. For the distance between destinations, residents have to adapt it passively without choice; for travel mode, residents could choose different ways to travel, depending on transportation system and individual demands and constraints. So how residents respond to the transportation environment in selection of travel mode is explained in this section.

The pattern of commuting of residents in the three SCHEs was different significantly, X^2 (6, $N = 384$) $= 31.717$, $p < 0.001$, Fig. 7.3 shows that in LNT, similar percentage of residents commuted to work by active travel mode (44.4%) and by public transportation (43.0%). In AG, most residents commuted by active travel mode (37.9%), followed by the public transportation (27.3%) and private car (22.4%). Most GV residents (44.3%) took public transport to commute, followed by those who took active travel (29.5%) and drove to work (23.9%).

For the public transportation environment, Table 7.5 summarizes the transit information around the three SCHEs. Without a metro station and uneven distributed bus stops, AG had the most inconvenient public transportation system. LNT had the most convenient one with 39 bus routes connecting other regions. GV had the most

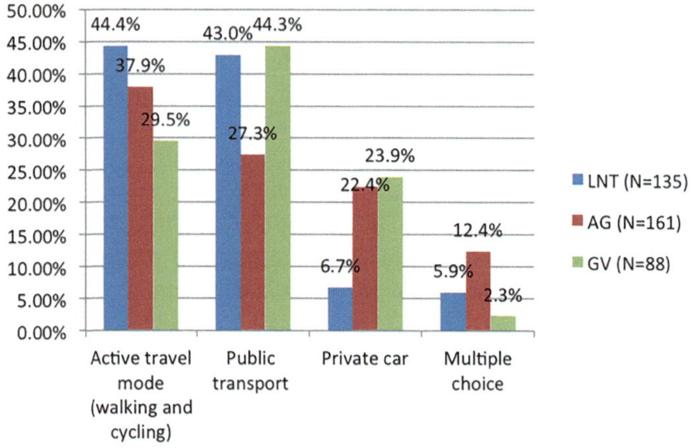

Fig. 7.3 Travel modes of commuting by case

even distributed bus stops, though it had the fewest bus routes. For walking environment, observations show that AG was at the advantages of the landscape avenue, and GV benefited from the green scenery along the streets, but both of them were at the disadvantages of single land use, decreasing destinations' accessibility. LNT was at the advantages of mixed and compacted land use but the visual quality of the streets was not high (Appendix 1). To examine how residents responded to the transportation environment, comments in interviews could be referred. Focusing on commuting, Table 7.6 unfolds residents' values, opinions, and responses towards their transportation environment.

Such comments suggest that three characteristics of the travel mode affect travelers' behaviors: comfort, convenience, and flexibility. Generally, the public transportation is not advantageous in terms of comfort and flexibility, since the environment of the coaches was crowded and smelly and it costs time to wait. Private car is much comfortable and flexible, especially with regards to standing among the crowded passengers and waiting long for certain buses. For convenience, characteristics of the transportation environment of certain cases are influential, such as that LNT had convenient public transportation system but AG had inconvenient one. Although driving is convenient for getting to the destinations, the difficulty of parking increases the inconvenience of using private car. Thus, it is proposed that residents' travel behaviors are the adaption to these three aspects of transportation environment.

Individual demands could affect their travel behaviors, which could be adjusted in order to adapt to the transportation environment. For example, Case 2 and Case 4 in Table 7.6 gave inverse comments on the comfort of the metro, and this is because that case 2 had to go to work in urban core area every day while case 4 only needed to travel to urban core area from time to time, so the former one was travelling in the peak hour while the latter one usually traveled in the off-peak hours. In addition, individual attributes also have influence on travel behaviors, such as the income showed in Case

Table 7.5 Public transportation environment by case

<table>
<tr><th>Graphic analysis</th><th>Remarks</th></tr>
<tr><td>

</td><td>LNT
1 subway station
8 bus stops
39 bus routes</td></tr>
<tr><td>

</td><td>AG
No subway station
5 bus stops
17 bus routes</td></tr>
</table>

(continued)

Table 7.5 (continued)

<table>
<tr><th>Graphic analysis</th><th>Remarks</th></tr>
<tr><td>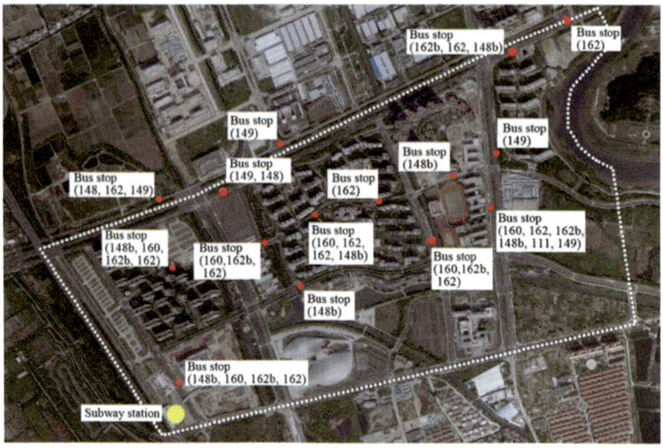
</td><td>GV
1 subway station
14 bus stops
7 bus routes</td></tr>
</table>

4 in Table 7.6. Thus, this study argues that the responsive transaction between travel behavior and transportation environment is a passive process for residents, through adjusting their demands and expectations.

7.1.2.2 Physical Leisure Activity: Preference, Place Chosen, Variety, Regularity

Previous analysis finds that the pattern of conducting physical leisure activity was associated with residents' health: the preference, place chosen, variety and regularity of doing physical leisure activity. Since the preference for physical leisure activity depends on individual personality greatly, this section focuses on the effects of the perceived leisure and design features on variety, regularity and place chosen of physical leisure activity.

Spearman correlation analyses were performed, and the results are listed in Table 7.7, showing that the variety of engaging in physical leisure activity was correlated with safety, quantity of open space, ambient environment, and sense of belonging. Regularity was correlated with quantity and quality of open space and fitness facilities, safety, density, ambient environment, and sense of aesthetic and belonging. However, no correlation was found between place chosen and the environmental factors. Thus, it is inferred that regularity and variety of doing physical leisure activity tend to be affected by the perceived residential environment, while the place chosen for physical leisure activity had not been affected by leisure and design features.

These findings are consistent with previous studies. Based on a review study on the relationship between physical environment and physical activity behaviors,

Table 7.6 Statements on travel mode for commuting

Statement	Choice	Cognitions for choice		
		Comfort	Convenience	Flexibility
1. I like to commute by my car, which is quite flexible and convenient. (GV-MM3)	Car		✓	✓
2. My children drive to work every day. They do not like to take the subway because there are too many people during the peak hour. It is very tired to stand for the whole journey. (GV-FO6)	Car	✓		
3. Subway is too crowded and uncomfortable. Since the distance is quite long from my home to work, I have to spend long time on the way, affecting my mood time to time. (LNT-FY1)	Subway	✓		
4. My children have to take subways for around one hour to go to work, which is really hard. But they couldn't afford a private car right now. (GV-FO4)	Subway		✓	
5. I use both public transportation and a private car to travel. It depends on the destination. If I traveled to the urban core area, taking subway would be more comfortable, because parking cars is troublesome. (LNT-FM4)	Subway or car	✓	✓	
6. For commuting, I either take buses or dive the car. It mainly depends on the convenience and comfort of the travel mode. If the bus could link my destination well, I prefer the bus because driving is tired. If it was too crowded in bus, I would choose to drive. (LNT-MM6)	Bus or car	✓	✓	
7. I either take the shuttle bus or drive the car. It is convenient and relaxed when the shuttle bus connects to my destination well. (AG-FY4)	Shuttle bus or car	✓	✓	
8. I like to use Uber to travel because it is flexible in terms of the time. (AG-FM1)	Uber			✓
9. I don't like to take public transports since the bus and taxi are very smelly and they cost much time to wait. Usually I choose Uber to travel, which is very convenient and comfortable. (LNT-FM2)	Uber	✓	✓	✓

(continued)

Table 7.6 (continued)

Statement	Choice	Cognitions for choice		
		Comfort	Convenience	Flexibility
10. I walk to my office every day, which only takes me 10 min. (LNT-FY7)	Walking	✓	✓	✓
11. I cycle to work with the aim of doing exercise. Although cycling could make me feel well, it is not very convenient in terms of the roads from my home to work are for cars not for bicycles. (AG-MY3)	Cycling	✓		

Table 7.7 Correlations between doing physical leisure activity and perceived residential environment factors

Perceived leisure and intangible environment	Variety of leisure activity		Regularity of leisure activity		Place chosen of leisure activity	
	r_s	sig	r_s	sig	r_s	sig
Leisure environment						
Quantity of fitness facilities	−0.040	0.363	−0.109*	0.014	0.048	0.366
Quantity of open space	0.037	0.396	0.113*	0.011	−0.017	0.751
Quality of space and facilities	0.008	0.865	0.144**	0.001	0.007	0.890
Intangible environment						
Safety	0.113*	0.010	0.084	0.056	0.014	0.795
Density	0.053	0.226	0.118**	0.007	0.027	0.611
Quantity of green space	−0.163***	0.000	−0.071	0.108	−0.003	0.960
Ambient environment	0.103*	0.019	0.265***	0.000	−0.020	0.706
Sense of aesthetic	0.076	0.085	0.180***	0.000	0.070	0.189
Sense of belonging	0.132**	0.003	0.283***	0.000	−0.047	0.374

Note * significant at $p < 0.05$ level, ** significant at $p < 0.01$ level, *** significant at $p < 0.001$ level

Humpel et al. [5] summarize five sets of environment factors associated with physical activity: safety, accessibility of facilities, opportunities for activity, and aesthetics. For the explanation of influencing mechanisms, most studies take it for granted that some environment settings such as sport fields, gyms, and bicycle trails could promote physical activity apparently [10]. So the possible causal relationships need to be identified [5].

Observations indicate the low utilization rate of the space and facilities for physical leisure activity. The photographs of the fitness facilities depicted in Fig. 7.4 were taken on workdays and on weekends, showing that few people used them. Table 7.8 summaries residents' opinions on the low utilization rate and their dissatisfaction.

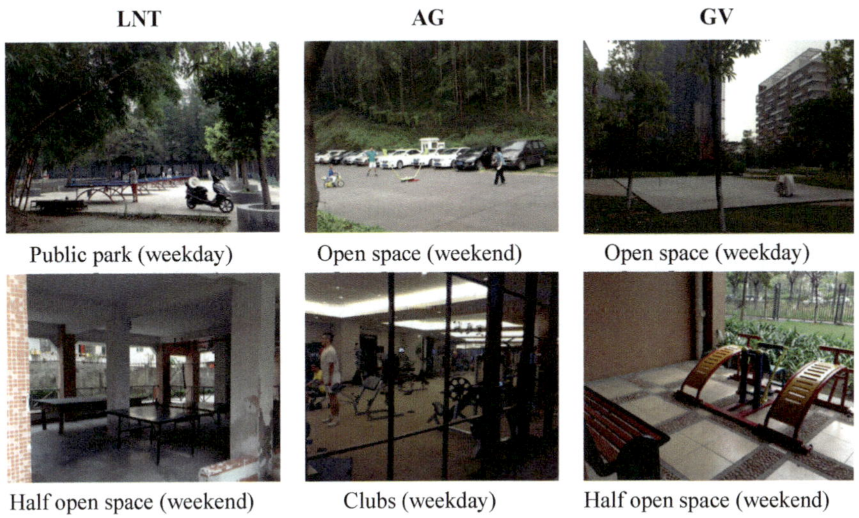

Fig. 7.4 Place for physical leisure activity by case

Such statements reveal that the young or middle-aged residents tended to believe that most fitness facilities were not for them (see Case 1, 2, 9 in Table 7.8). However, the elderly seldom used the fitness facilities (Case 8) or they reported inadequate facilities for them (Case 7). Both younger and older adults complained that the fitness facilities failed to meet their needs, and the demands of the elderly for fitness facilities were misunderstood by residents. Thus, the mismatch between the leisure environment and individual demands in both quantity and variety could be an essential reason for the low utilization rate. In this sense, residents had to adapt the mismatch in a passive way such as choosing to do physical leisure activity outside of the estate.

In addition, focusing on the effects of environment on physical leisure activity, residents' comments in interviews indicate the responsive transaction directly.

> The open space in our neighborhood is quite small, and there is no large open space for doing exercise. There is only a platform on third floor, where located some fitness facilities for children. Few adults do exercise there. So most adults and the elderly go to the riverside. (LNT-FM4)
>
> The fitness facilities in our residential cluster are not enough, but we would find alternative ways such as taking children to other residential clusters to use the facilities there. This is the advantage of the large-scale neighborhood. (AG-FM1)
>
> If there were more available and accessible fitness facilities or parks, I would do more physical exercise. Especially, if there were some green spaces near my home, I would walk more. The lack of facilities makes it is impossible for people to do physical leisure activity. (LNT-FM5)
>
> I think people are very easily induced by the environment, both social environment and physical environment. People need some motivations to do exercise, either from motivations from friends or from the attraction of the fitness facilities and parks. (LNT-MY9)

Table 7.8 Statements on fitness facilities in the three cases

Statements	Opinions	Age
1. There are some fitness facilities in the public space, but most of them are for children, so adults and the elderly usually go to the riverside to exercise. (LNT-FM5)	For children	Middle aged
2. The fitness facilities only appeared in our neighborhood recently; there were only stone seats and children's slide. (LNT-FY7)	For children, monotonous type	Young
3. If the neighborhood could provide more public cultural or sport platforms or facilities, it would be very good for the health of elderly since they have nothing to do except babysit (LNT-MY9)	Lack of facilities for the elderly	Young
4. I rarely use the fitness facilities, but it is important for improving the quality of life. Although I do not use it, it is necessary to have choices. (AG-FM1)	Don't use, but need the various opportunities	Middle aged
5. There was a football field before, but now we do not have it anymore. Although there are courts for ballgames, I still desire to have the track for running, which is totally different from the running machine. (AG-MY3)	Ask for more facilities	Young
6. I rarely use the fitness facilities, and I usually walk around and practice Tai Chi. (AG-MO5)	Do not use	Older adult
7. There are inadequate fitness facilities. There is nothing appropriate for us to use. (GV-FO1)	Inadequate	Older adult
8. There are a lot of fitness facilities, but very few people use them. (GV-MO5)	Do not use	Older adult
9. I think the fitness facilities are all right, and most of them are for the elderly and children. (GV-FM2)	For the elderly and children	Middle aged

These comments suggest that residents have to adjust frequency and variety of physical leisure activity according to the constraints of the leisure environment, such as the lack of the open space and fitness facilities. Conversely, residents also adjusted their physical leisure activity pattern according to the activity-promoting features of the residential environment, such as the possible effects of attractive open and green space extending walking time. In addition, residents may be influenced by the behaviors of people around them, and that the atmospheres of conducing physical leisure activities within the neighborhood could raise the motivations of doing exercises. Therefore, this study finds that residents tend to passively adapt to both constraints and opportunities for physical leisure activity in a neighborhood through adjusting the variety and frequency of their engagement. Motivation of doing physical leisure activity could mediate the responsive transaction process.

To summarize, this study argues that the pattern of physical leisure activity is the result of passive responsive transaction. Residents adapted to the residential environment (both constraints and opportunities) by choosing other places outside the neighborhood to do physical leisure activity, adjusting the variety and regularity of participation.

These comments suggest that, first, residents have to adjust frequency and variety of physical leisure activity according to the constraints of leisure environment, such as the lack of the open space and fitness facilities. Conversely, residents also adjusted their physical leisure activity pattern according to the activity-promoting features of the residential environment, such as that attractive open and green space could extend the walking time. In addition, residents may be influenced by the behaviors of the surrounding people, and the atmospheres of conducing physical leisure activities within the neighborhood could increase the motivations of doing exercises for individual. Therefore, this study finds that residents tend to passively adapt to both constraints and opportunities for physical leisure activity in neighborhood through adjusting the variety and frequency of their engagement. Motivation of doing physical leisure activity could mediate the responsive transaction process.

To summarize, this study argues that the pattern of physical leisure activity is the result of passive responsive transaction. Residents adapted to the residential environment (both constraints and opportunities) by choosing other places outside the neighborhood to do physical leisure activity, adjusting the variety and regularity of the participation.

7.1.3 Operative Transaction: Participation in Housing Operation

Operative transaction refers to the direct impacts of individual's movements on the environment [14]. In suburban living, participation in the operation of the housing environment is the main way for residents to modify their living environment, which is the focus of this section. Although many studies associate residents' participation in local decision-making and the participation in planning process with health outcomes [4, 15], this study finds no significant associations between participation in housing management and the risks of mental disorders of residents in SCHEs. To see whether residents' participation in housing management affairs affect the residential environment, two arguments are examined based on the interviews.

It is assumed that residents' participation has little effect on the physical residential environment. One reason is the participation was quite low, Fig. 7.6 suggests that as many as 57.9% of residents did not participate in housing affairs. Another 24.2% discussed operations with neighbors; only 15.8% appealed to the housing manager or residents' committee to talk about housing management affairs. Although some residents expressed that they would like to be a member of homeowner association (4.8%) or establish the self-organized organizations (4.2%) or attend the self-organized community initiatives (3.2%), there were no such organizations in all the three cases yet (Fig. 7.5). In addition, the frequency of actual participation was very low: only 17.9% respondents participated at least once a month (Fig. 7.6). Thus, based on the low degree and low frequency of participation in housing management affairs, statistically residents' participation in the housing management affairs had very limited influence on residents' health. Interviews also supported this argument.

> Many green spaces between housing blocks have been transformed into parking lots with the increasing numbers of private cars, which makes us feel disappointed and depressed. Although we expressed our dissatisfaction to the housing managers, they could do very little. With time, we just accepted it since you have to park somewhere. (AG-FM1)
>
> The housing manager is more interested in earning money than in providing good service. The maintenance of facilities is poor in this neighborhood, but complaining to the housing manager makes little difference in the daily operation. We have to put up with for this situation. (LNT-FM5)
>
> The housing management in our neighborhood is quite poor. For example, the light in the staircase has been burned out for several months, although we reported this to the housing manager, it still hasn't been replaced. We have to put up with it. (LNT-FM4)

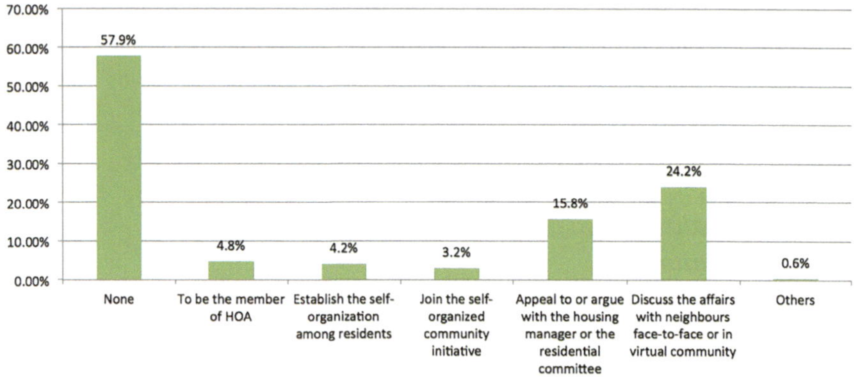

Fig. 7.5 Types of participation in neighborhood operation ($N = 525$, Response = 670)

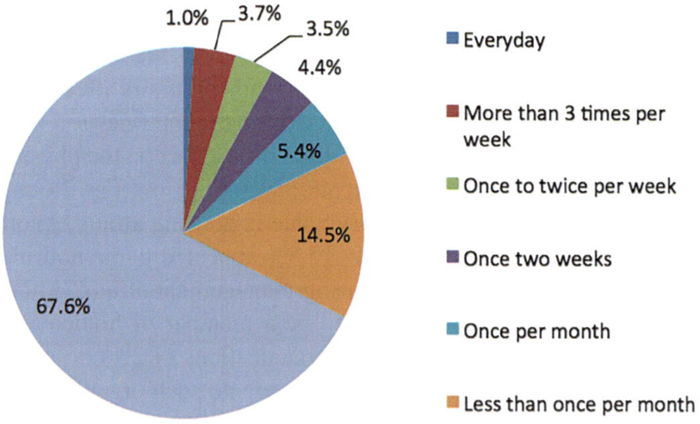

Fig. 7.6 Frequency of participation in neighborhood operation ($N = 519$)

Such attitudes displayed either frustration or helplessness in terms of participation in housing management affairs. The popular way of participation is reporting problems to housing manager. Residents had to put up with the lasting conflicts between residents and housing manager in terms of the management services. In this sense, there is no way for residents to modify their living environment.

However, residents' participation in the housing management affairs has some influence on social network and neighborhood cohesion. Many residents discuss the operational issues with their neighbors, which may increase the interactions with neighbors and build neighborhood cohesion. Two residents remarked that:

> We often talked about the issues related to the housing management, such as the issues of building quality. Sometimes, these discussions could increase the attachment among neighbors. (GV-MM3)
>
> Neighbors share their issues related to the housing quality to each other, and complain about the service provided by the housing manager. Sometimes, this kind of complain makes people know each other, increasing the connections among neighbors inexpertly. (AG-FM1)

Thus, although residents' participation has little influence on the improvement of physical residential environment, it does reinforce social networks and neighborhood cohesion.

Given the low rate of participation, it could be hypothesized that residents do not have the initiative to participate in housing management affairs. Some specific reasons were given in the interviews:

> Participation or not could be influenced by the individual's value and occupation and whether the issue is highly related to the individual's interests. For example, residents may not participate in the fair that the green space is transferred into parking space unless the green space is located around their own home. (AG_MY7)
>
> I think many residents do not like to participate the housing operation since they are busy earning money and it is not a benefit to participate. Only those with time, money and passion would participate in operational affairs. I believe most middle-aged people and young people would not participate due to their limited time and energy. (GV_MY3)
>
> Most residents have no time or interests to take part in housing operation unless it has something to do with their interests. For example, if a waste incineration plant was planned to be built near the neighborhood, I believe most residents would express disagreement. (AG_MY3)
>
> Although we want to change something to improve the residential environment, there is no way to do so in reality. Even for small leaks in the apartment, we have to fight with the housing manager with wits and courage, let alone the public issues. It is too hard and tiresome to deal with operational issues with the housing manager. (LNT_FY7)

Such comments suggest that time constraints, the lack of impetus, and channels for participation diminishes residents' participation in housing management, and hence reduces their influence in improving the physical residential environment. From the good side, it is inferred that there were no serious public issues and interests' conflicts in SCHEs that have adverse effects on residents' well-being. In short, the operative transaction is generally absent in residents-environment transaction of SCHEs, while its negative effects are moderate.

Based on the examination of the interactions between environmental and behavioral factors, this study ascertains that the residents-environment transaction mode of living in SCHEs is a passive adaption one. The physical residential environment influences residents' cognition, which indirectly determine the responses of residents in the form of physical activity, and in turn, health. Furthermore, the factors influencing health could be traced back to the physical residential environment in the form of two groups of factors: public service facilities and design figures. The next section explores these factors that underlie residents' health.

7.2 Institutional Factors in Suburban Housing Development

The second part of this chapter discusses the institutional factors of suburban housing development that affect the formation of the residential environment. To identify the primary determinants of the physical residential environment of the SCHEs, the roles of the state, the market and the society, and their interactions in housing development are examined, focusing on public service facilities and design features.

7.2.1 Initiatives of Suburban Housing Development: The Role of Hierarchical Governments

Two initiatives of suburban housing development merit attention: suburbanization and housing commodification. The role of the state is crucial to both. In socialist China, housing is driven by state policy [2]. Although the housing reform helped to establish the markets in land, housing, capital, and labor, the state continues to play a crucial role [20]. The changing role of the state could be identified in housing development and suburban development.

After the housing reform, the role of state in housing delivery has transformed from direct intervention to making policy frameworks, implying an enabling and strategic role of steering and arranging housing services [23, 24]. In housing development, the state not only acts as the builder of market [25], but also shares power with market and society by empowering work units and citizens, making them stakeholders [24]. At the same time, a new and distinctive round of suburban development emerged in the late 1990s [13, 20]. Initiated by central cities' thirst for land and other production factors in response to regional and global competition, many municipal governments underwent administrative annexation to expand their space and resources, gaining power and control over across the metropolitan area [20, 21]. Dual initiatives for suburban development—state entrepreneurialism and growth machine—could arguably be identified under the context of urban governance transformation in China [20]. These initiatives will be elaborated by the development of the three SCHEs.

Administrative adjustment is an important contextual factor for suburban housing development in Panyu, implying the effects of the power relations at different levels of governments on the formation of the unique characteristics of suburban housing estates. As discussed in Chap. 3, there were two administrative adjustments respectively in 1992 and 2000, transforming Panyu from a county into a county-level city and then into a district of Guangzhou. Relations between different levels of governments have grate influences on the housing development in Panyu. In the following discussion, due to the changes in administrative jurisdiction, the term "county government" refers to the government of Panyu city before 2000, and the term "district government" after 2000; the term "Guangzhou municipal government" is for the Guangzhou government after 2000.

Before 2000, although Panyu was administered by Guangzhou government due to its lower administrative hierarchy, the financial management of Panyu was independent [22]. The thirst for the capital for urban construction in 1980s forced Panyu county government to transfer the land use rights and spatial resources with capital [22]. Under this context, Panyu county government collaborated with Hong Kong companies to develop housing projects in the form of real estate joint companies. LNT was developed under the state-market cooperation.

In the late 1990s, fearful of losing its independent financial power with the administrative annexation in 2000, Panyu county government leased many construction land parcels to real estate developers through land lease agreements, greatly increasing the county revenues [22]. This initiative resulted in land use transformation from the previous planned industrial land to residential land, which brought great profits as fiscal revenue for Panyu county government shortly before the administrative adjustment, something that the Guangzhou government failed to predict and control. Thus, the interest conflict between Panyu county government and Guangzhou government has led to the booming of commodity housing estates in Panyu [22]. As a result, real estate developers gained huge parcels of land from late 1990s–2003, paying as little as 40,000–80,000 yuan per unit of area (mu).[1] In addition, the twelve largest housing estates have taken up 15.63 km^2 in Panyu [22]. AG was developed under these conditions.

With Panyu turned into a district of Guangzhou in 2000, Guangzhou municipal government has recaptured both fiscal power and approval rights for land leasing and construction projects in Panyu. There were consequences on land and spatial development, the first of which was that Guangzhou municipal government enhanced the regulation of construction land leasing in Panyu to prevent the land enclosure of the developers, caused by the large land transfer from the late 1990s to the beginning of 2000. The second consequence was that since most essential land resources had become the control of developers in the northern part of Panyu, Guangzhou municipal government had to revise the strategic planning of Guangzhou metropolitan

[1] Retrieved from http://www.time-weekly.com/html/20141014/26732_1.html; 1 km^2 = 1500 mu.

area, highlighting the new axis in Panyu which comprised the Olympic center, Pazhou exhibition center, University town, and Guangzhou new town. In these areas, land for industrial development was plentiful [22]. Under this context, new commodity housing development in Panyu has been greatly influenced by the desire of Guangzhou municipal government for revenue from land and spatial development. For example, taking the opportunity of Asian Game in 2010, the athletes' village was located in Guangzhou new town (GV), reflecting the desire of Guangzhou municipal government to encourage urban spatial growth. The development strategy was that "state dominated and participation with the market", meaning that state is responsible for planning regulation and the construction of the key infrastructure and public facilities, while the market-controlled land development and urban operation [17].

To summarize, competitive interests and changing power relations within the government hierarchy are of great importance in housing development in Panyu, especially in terms of land leasing process and the location of the housing estates. In addition, the dynamic relations between the state and the market in suburban housing development are affected largely by the manipulation of spatial and land resources and capital resource.

7.2.2 Decision Making of Suburban Housing Development

The developer dominates decision making in the development process of suburban commodity housing from inception, planning and design to construction, shaping the characters of the residential environment. Focusing on the decision-making process of housing development, this section examines the roles of developers and its interactions with local governments and planners. Two aspects of property development that have significant effects on residents' health are investigated: neighborhood types and public service facilities.

7.2.2.1 Neighborhood Types in Suburbs

It is argued that market conditions determine the type of housing products that developers desire to provide. Developers usually make decisions on desirable projects according to the market conditions, such as the supply and demand situations of properties and capital. Four major considerations for determining the nature of housing projects emerge from interviews with several developers in Guangzhou: (1) location and advantage in resources, (2) target population, (3) regulatory environment, and (4) the development stage of the developer itself.

First, location and surrounding resources (natural resources, infrastructure and public facilities) could determine the best type of housing to be developed. Two developers mentioned that the features of the land parcels are essential considerations.

Deciding the orientation of a housing project depends on location and existing resources of the site. Compared with housing projects in urban core area, suburban housing should be developed according to its scarcity in terms of the attributes such as the affordability, good natural environment, and the possibility of developing villas. (Developer B, 02-12-2015)

The characters of the land site determine what kind of product you could provide, especially in terms of the site scale, the location, the natural resources such as mountain and river. (Developer C, 02-12-2015)

Second, the target population has a direct effect on housing types and on the character of the residential environment. Initially, the decision of the target population depends on both market conditions and the policy environment. As Developer B stated, "to decide the target population, it is important to see which type of product could be sold more easily and bring more profits." This indicates that the housing demand and supply balance is at the center of developers' considerations in terms of target population. For housing in suburbs, three types of target population and their disparate demands are identified through interviews. They are first-time homebuyers, upgraders, and the extremely affluent, as the developers noted:

First-homebuyers, running after the life, caring much about the connectivity between home and work place, and the basic living standard could satisfy their demands. (Developer D, 03-12-2015)

Upgraders, middle class, who care much about the education of the next generation and the living environment of the neighborhoods, require much on the neighborhood relationships and pay more attention to the community activities than the first-homebuyers, have more requirements from the property management services aside from basic security. (Developer D, 03-12-2015)

The extremely affluent, who have more than one property, they have high requirements for the physical environment, especially the service facilities that could improve the quality of life. (Developer D, 03-12-2015)

Generally, upgraders and the affluent population choose living in suburbs for the good environment, while the first-time homebuyers choose suburban housing for its affordable price. (Developer B, 02-12-2015)

The purchase motive is very different from the first-time homebuyers, upgraders to the affluent people, resulting in the various demands. Payment ability is a crucial factor influencing the demands of consumers. (Developer B, 02-12-2015)

Such comments suggest that different target populations have different purchase motive, specific demands and constraints. The aim of subdivision of the consumers is to identify the different demands, which result in different planning and design of the residential environment. Accordingly, three types of suburban housing have been identified: housing for first-time homebuyers, housing for upgraders, and luxury housing. Table 7.9 lists the difference among the three types of housing.

Thus, first-time homebuyers are more likely to be attracted by affordability and convenient transportation; upgraders and the extremely affluent prefer spacious room, good environmental quality and additional services. Comments of developers point out the effects of consumers' demands on housing planning and design. As two developers mentioned:

Table 7.9 Three types of suburban housing

Requirements	Housing for rigid demand	Housing for upgraders	Luxury housing
Housing layout	One bedroom; Two bedrooms	Three bedrooms	More than three bedrooms; Deluxe suite
Green space and landscape	No requirement	Basic requirement	High requirement
Support facilities	Convenient transportation	Good educational facilities	High requirements in terms of the educational, healthcare facilities and recreational facilities
Housing price	Affordable	Moderate	Expensive

Source Organized by author based on the interview with developers

> It is necessary to subdivide the consumers to identify and predict the demands of different consumers precisely, which could be helpful in terms of project positioning and development strategy. (Developer A, 02-12-2015)
>
> Housing for first-time homebuyer are the easiest to produce, because most first-time homebuyers have similar demands. Technically, the plot ratio of first-time homebuyers' housing is higher than those for upgraders and there will be more rooms facing north in the housing for rigid demand. (Developer C, 02-12-2015)
>
> The housing for upgraders usually emphasizes the good living environment, such as the large garden, the good view of mountain and water body. In addition, developers usually provide more green space for upgraders than for first-time homebuyers, which cost more than merely provide the hard paving squares. (Developer A, 02-12- 2015)

Thus, the selection of target population is determinative for the housing type and housing planning and design.

Third, the macro policy regulation for housing market conducted by the government is also influential to housing types. For example, influenced by the 2011 restrictions on housing purchase, the trading volume of the luxury housing has decreased, leading to the increasing provision of housing for first-time homebuyers. Developers paid more attention to satisfying the rigid housing demand instead of developing luxury houses. Under the 2014 policy of relaxing the down payment on mortgages for second home in Guangzhou, the housing for upgraders has attracts the attention of the developers. Thus, the housing policies and planning regulations conducted by governments have great influences on housing types.

Forth, the stress of recouping investment is quite influential in determining the types of housing products, especially for the startup real estate enterprise. As Developer C mentioned: "if the aim of one housing project is to recoup fund quickly, developer may not focus on the consumers that bring highly added value but on the consumers who could reduce the period of fund return."

To conclude, the types of housing projects in suburban area are determined by natural resources, market conditions and government regulations, with developers' purposes of making profits. For the long-lasting housing projects, such as LNT,

AG and GV, developers usually base their choice of housing types on the dynamic housing market by mixing different housing types, suiting the demand and lifestyles of different housing consumers.

7.2.2.2 Service Facilities

In terms of planning and design of service facilities, the interplays between the developers and the local government, and between the developers and the planners are examined.

Relations between the developers and the local government

The transportation service and public facilities for the housing estates fall under the auspices of public affairs. In suburban housing development, relations between the state and the market are influential in the provision of service facilities. As a result of leasing a large volume of cheap land around 2000, large parcels of residential land in Panyu were transferred to the control of developers, and Panyu county government lacked the capital for infrastructure and public facilities. Under these conditions, developers had to provide both infrastructure and public facilities to attract home-buyers, such as traffic facilities, schools, hospitals, and daily commercial facilities. Thus, decision-making authority for the service facilities belonged to the developers. For example, the AG developer built the main road linking the Panyu Avenue and AG. The developer also provided the shuttle bus linking the urban core area and AG to compensate for the shortage of public transportation. However, the developers-depended provision of service facilities is problematic, leading to many issues such as the lack of public transportation and insufficient educational and healthcare resources. These problems cannot be solved by profit-driven developers alone.

With stronger planning regulation after 2005, Guangzhou municipal government gained more regulatory power over the public service than the previous stage through withdrawing of the approve rights of Panyu district government and the launch of the regulation plan throughout the whole city of Guangzhou. This change has been observed in the development of AG, and a statement given by Official A from the planning department shows that the regulations have been tightened.

> In AG, the development of the residential clusters after 2005 or 2006 was required by the municipal government to follow relative strict planning regulation. AG has to follow the mode of providing public facilities in Guangzhou, which require the developer to construct the facilities first and then transferring the ownership to the government after completion. (Official A, 04-12-2015)

In GV, transportation-oriented development (TOD), environment-oriented development (EOD), and service-oriented development (SOD) have been used with the approval of Guangzhou municipal government. Accordingly, GV has taken advantages of the government's regulation, especially the construction of a metro station, the key middle school, and the A-level hospital. However, constrained by limited capital, Guangzhou municipal government assigned the whole project to a joint real estate company, Lihe Property, which is run by five prominent developers: Agile Property, R&F Property, Country Garden, Shimao Property, and Citic Real Estate. So the developers were able to manipulate the development under the policy framework approved by Guangzhou municipal government. This led to unsatisfactory implementation of the planning and design, such as the peripheral location of the transportation and public facilities and poor service for daily necessities, especially the retails. Interviews highlight two possible reasons for the failure in providing commercial facilities for daily necessities. One was that GV was leapfrogged in a large parcel of farmland, which makes it is difficult to attract enterprises in a short period. In addition, it was speculated that the transfer of development authority to the developers makes it possible for them to decide when to provide the commercial facilities according to the market conditions but not residents' demands. From the start of 2016, the developer finally began to build commercial facilities.

It can thus be argued that either the previous solely or partially dependence on developers has caused issues in public service provision. However, the comparison of AG and GV suggests that the macro intervention of the local government could benefit the provision of public transportation and public facilities. Thus, it is argued that the enhanced planning regulation is beneficial for the health-promoting residential environment either intentionally or unintentionally.

Relations between the planners and the developers

The input of planners could not be neglected. However, their planning and design decisions have to be approved by developers who hired them. As demonstrated above, whether the developers would like to spend money on the provision of additional service facilities depends on expected economic benefits. What this means is that although the planners may be concerned with the health of residents, the same might not be true of the developers. In addition, planning and design decisions are affected by the governments' regulations through land leasing conditions, regulation plan, and approval of the permits for the housing project. Interviews with planners elicited the following insights:

> Frankly, we planners and designers belong to the service business, providing professional service for our customers. For the housing project, most customers are the developers, whose value orientations are influential to our plans and designs. Although we prefer the health-oriented design, it is hard to implement with many constraints from the developers. (Planner B, 30-11-2015)
>
> Our decision autonomy is very limited, especially in terms of what types of facilities and buildings should be provided. We are constrained by the developers and the planning conditions given by the government most of the time. However, we can try our best to adopt the people-oriented principles in our design under these constraints. For example, we usually promote the environmental equity and try best to balance the distribution of good resources. (Planner E, 30-11-2015)
>
> The orientation of the housing project affects the planning and design of public services facilities. This is determined by developers' consideration according to the market conditions, especially the target population they want to attract. We planners and designers usually have to follow their orientations. (Planner C, 30-11-2015)

Such opinions indicate the subordinate position of the planners, the decision role of developers, and the regulatory effects of the government in terms of basic decisions about planning and design for service facilities. It is inferred that the government regulations guarantee the basic standard of public service, and developers meet the requirements through the efforts of the planners. From the lens of developers, market conditions again are determinative in their decisions on service facilities. As one planner and one developer pointed out:

> A major considerations of the developers is the cost, which is the constraint for them in providing high quality public service facilities. The final aim of developers is earning more money. (Planner F, 30-11-2015)
>
> Actually, what facilities will be provided in a housing project depends on the market. For example, if the consumers are willing to pay more on an additional basketball court, the developers will build more basketball courts to earn more money. (Developer A, 30-11-2015)
>
> Sometimes, developers package their products to attract potential buyers by providing additional facilities or space. For example, although first-time homebuyers are sensitive to affordability, the services for children and space for physical activity are preferred by residents, especially when the housing market is not well. (Planner D, 30-11-2015)

Theses opinions confirm that the planning and design of service facilities depended on the developers' orientation, which is shaped by market conditions. So market conditions are the basis of developers' decisions on the amenities added to housing estates. The demands of consumers, cost–benefit balance, and expected profits could be the main consideration of the developers in the planning and design of service facilities. Thus, this study confirms that the developers are only interested in consumers' demands that could promise economic benefits, not in people-oriented and health-promoting services.

7.2.3 Institutional Factors Influencing Health-Promoting Environments

Referring to the examination of initiatives and decision-making of suburban housing development, the institutional factors influencing the health-promoting environments could be proposed. The analysis indicates that there is no health-promoting orientation in suburban housing development, neither from the perspective of the state nor from the market. This may partly be due to the ignorance of health-promoting environment in China, or to the incapability of the current governance for suburban housing development, which cannot protect health as a public interest.

7.2.3.1 Intuitional Uncertainty and Establishment of Neoliberalism in China

Institutional uncertainty in the development of suburban housing has impeded the implementation of health-promoting factors, such as comfortable and convenient public transportation, the affordable and high-quality schools, satisfactory commercial services, ample opportunities for physical leisure activities, and plenty of green space and strong sense of belonging.

Institutional uncertainty is generated when "the institutions of the status quo are weakened and new institutions are yet to be established" [26]. During China's economic reforms and social transformation since 1978, there has been considerable institutional uncertainty. Affected by economic reform and opening policy, Chinese government has been transformed from a socialist welfare provider to an advocate for marketization, with powering the market as a provider [26]. The roles of the governments and the market have changed dramatically, contributing to the transition from central planning to market orientation [26]. However, although the centrally controlled system has been weakened, the market economy has not been fully established in China. This indicates the coexistence of central planning and commercialization, leading to institutional uncertainty that could affect commodity housing development [26].

Institutional uncertainty could result in the ambiguous and incomplete land rights (especially over the collectively owner land), the disorganized short-term behaviors such as disorderly land rent competition and informal development, and the fragmented land development mingling both top-down projects and bottom-up constructions [26]. Referring to the three periods of SCHEs development in Guangzhou, there are transitions from market-government collaborated, market-dominated to the regulation-enhanced periods. Dynamic relations between the governments and the developers accelerate the institutional uncertainty in commodity housing development, contributing to the emerging issues of SCHEs development. Institutional uncertainty could be a fundamental reason for the speedy and volumetric land leases to developers around 2000, the developer-depended provision of public service facilities, and the leapfrogged housing development led by top-down projects of Asian Game Village and Guangzhou New Town.

Furthermore, institution uncertainty is associated with the relationships across state, market and society, which are influenced largely by the establishment of neoliberalism in China. The neo-liberal model is a popular theoretical approach for understanding housing development in contemporary China, emphasizing weakened state intervention and superiority of market processes [18] (Fig. 7.7). Neo-liberal reforms brought significant changes to the provision of urban housing in China, establishing the housing market. In commodity housing development, the governments focus on the creation of financial and institutional environment for housing market and the regulation of market performance [18]. In addition, maintaining political and social stability and consolidating the rule of the Communist Party are also important purpose of the employment of neo-liberal housing policies [18]. In addition, neo-liberal reforms lead to the shift to market-based governance, contributing to various policies for market discipline and deregulation [12]. So social and economic affairs increasingly rely on private investment, and there is a higher priority on private interests than public benefits [12], as cited in Brenner and Theodoore 2002). Moreover, the state-market relation is distinctive in China, such as what Wu [19] notes:

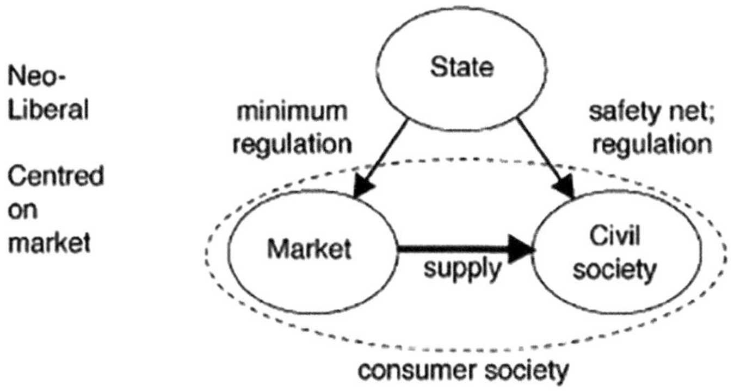

Fig. 7.7 Roles of state, market and society

"Chinese neoliberal model is comprised of two seemingly contradictory yet complementary elements: a market mechanism and strong state control." Thus, with reduced intervention of government but strong will for global competition, in addition to the dominance of the market mechanism, it is argued that commodity housing developed under the neoliberalism is unlikely to protect the public interest: human health.

The above argument can be supported by the health literature. In the era of neoliberalisation, human development could be adversely affected by the neo-liberal development strategies, especially on the aspects of poverty, inequality, and health [9]. Navarro [9] reviews the effects of neoliberal economic policies on the health statuses of people in developed and developing countries, suggesting the deleterious effects of neoliberalism on the health of populations and the consequences for health system spending. For example, although some international organizations use the flag of human development, their initiatives under the context of global neoliberalism could be counter-productive and health-inequitable[2] [9]. To summarize, institutional uncertainty in the three periods of suburban housing development and the establishment and maturation of neoliberalism in China's housing market have created the institutional environment for developing the suburban housing estates. As this type of housing development is commercially oriented, its physical environment may not be health-promoting.

7.2.3.2 Institutional Factors Affecting Provision of Service Facilities

In neoliberal housing market, public interests such as health are not the priority of either the state or the market. As a factor affecting residents' health greatly, provision of service facilities is vulnerable to the institutional environment. This section aims to identify the institutional factors that affect the service environment.

[2] Retrieved from https://scholar.google.com.hk/scholar?q=neoliberal++and+health+&btnG=&hl=en&as_sdt=0%2C5.

The strategy of delivering the infrastructure and public facilities in Guangzhou is "developer construction, transferring to the government", which is an effective way of constructing service facilities, especially when the governments cannot afford to build them. In those cases, developers have to construct service facilities for the governments, and then hand them over to local government agencies, such as the Education Bureau, Electricity Authority, and City Management Committee. Regulations for Public Service Facilities in Residential Area in Guangzhou (revised in 2014)[3] note that service facilities of residential area should be constructed by the developer. There are three types of facilities based on the transfer of property rights: (1) facilities that should be transferred to the governments free of charge (such as education, community healthcare, cultural and recreational facilities, administrative facilities, and public utilities); (2) facilities that could be transferred at cost (such as transformer substations and postal stations); (3) facilities that do not need to be transferred and can be used by the construction company or the property owners (such as commercial facilities, parking facilities and gas station, cultural and recreational center, public square and green space). In operation, the provision of service facilities in SCHEs is regulated by Guangzhou municipal government through the planning and land lease conditions and planning approvals. However, this mode of providing public services in SCHEs is problematic, especially in terms of the transfers of property right, going against the public interest.

For the facilities that should be transferred to the government free of charge, the planning and design of the facilities is influenced by the transfer requirement greatly. As a result, these facilities are planned in a way that is easy for the property rights to be transferred. The allocation of independent land plots for the facilities has been promoted, which lacks an environmentally integrated plan, leading to poor quality of the built environment. Moreover, developers tend to adopt the most cost-effective way to build these facilities because they are neither the owner nor the user, which may result in the unsatisfactory built quality. In this context, although the regulatory plan has controlled the quantity of the facilities, it cannot control their quality and the completion status. The officials and planners who were interviewed offered explanations:

> It is like a wrestling match between the government and the developer, the former wants more, while the later wants less. This leads to the simplified implementation delivered by the developer with aim of reducing cost and maximizing profits. Thus, the completed facilities are unsatisfactory. (Official A, 04-12-2015)
>
> Developers prefer to arrange the facilities in a way that facilitates the transfer, and the negative effects of this initiative are aggravated by the government departments that take over the property rights of the facilities. Because of the need to transfer, the developers have no motivation to do their best in terms of the service facilities. (Planner A, 30-11-2015)
>
> The property right transfer causes the shortage of schools and hospitals in Panyu. Due to the unique administrative system of Panyu, the town governments (the lower layer government of Panyu district government) are the official body to whom the property rights of the service facilities should be transferred. However, the town governments in Panyu are unable

[3] Retrieved from Source: http://www.gz.gov.cn/sofpro/gecs/addidea_opinion.gecs?opinion.opinionSeq=7716.

to take over the public service facilities due to limited fiscal capability. In this case, the town governments postpone the transfer, leading to embezzling of the facilities by the developers. Taking education facilities in AG for example, the developers convert the public schools to private ones to earn more money because of the nonfeasance of the town government. (Official A, 04-12-2015)

These comments reveal that the problems of free-charge property rights transfer of the service facilities are stemmed from the stages of planning and design to construction. The governments are defaulted in term of the facilities' quality. The related governments' departments are irresponsible in taking over the property rights with the constraints of fiscal capability. This issue is particularly serious in educational facilities of SCHEs. As Official A pointed out, the lower-level government may not be capable to afford the operation of additional schools, so it leaves the schools to developers until it has enough fiscal capability. In this context, developers choose to develop private schools to attract consumers and to make profits, resulting in the lack of quality public schools in suburbs [6, 7].

When it comes to the facilities that do not need to be transferred to the government, developers seek more economic benefits. Taking daily commercial facilities for example, with the property rights, developers usually bring in the type of commercial activities that generate high rental yield, leading to the lack of the commercial service like barbershops that have low rental yield but that are needed. In the word of one official, "the developer of GV may wait for the time that could bring more profits than act right now to attract the investment in commercial activities." In addition, to capture the market, the developers pay more attention to planning, design and operation of the commercial facilities with the dynamic market conditions. Previously, such as in the AG and LNT, commercial facilities are found on the ground floor of the residential buildings. Nowadays, realizing the importance of the collocation of the commercial activities for raising the property price and improving the attractiveness of the housing project, developers begin to emphasize the centralized distribution of the commercial facilities and the rational collocation of commercial activities. As several planners remarked:

I think the neighborhood center is derived from development of suburban housing estates, and how to collocate different commercial activities become a topic that is deserve to study for the developers. (Planner B, 30-11-2015)

Developers pay more attention to the commercial facilities nowadays, which is planned with some public facilities together, forming a neighborhood center. For example, in a housing project of Vanke, the developer holds some commercial property and introduces some commercial activities with low rental yield (such as banks and postal service) into its neighborhood center. This measure increases the convenience of residential living. (Planner E, 30-11-2015)

Such opinions indicate that developers have impetus to improve the planning, design and operation of commercial facilities, because these facilities are beneficial for the developers in attracting potential buyers. In addition, sometimes developers provide facilities beyond the government requirements to allure their target consumers, especially the green space and landscape, and the service facilities. As two planers mentioned:

> For large-scale development, developers usually choose to finish the service facilities such as the shuttle bus and the commercial facilities before opening to sell, showing the potential consumers that the neighborhood could support the necessity of daily living and everything has been operated well. (Planner B, 30-11-2015)
>
> Aiming to promote the sales, developers usually provide the services or facilities that are attractive to the consumers such as large gardens and well-built service facilities, which are consistent with the interests of residents. (Planner C, 30-11-2015)

Thus, although the initiative of developers in provision service facilities and preferable amenities is profit-driven but due to the sovereignty of consumers, a people-oriented approach is increasingly adopted, including the health needs of the residents. In this sense, the economic benefits of developers and the interests of residents are consistent, which is somewhat beneficial to residents' health.

However, within the institutional framework pertinent to urban development, the government's regulatory authority becomes quite strong nowadays, guaranteeing the basic requirements of human health through regulation plan and administrative approval. Interviews with officials support this point:

> The control power of the government has been increasing over the years. During the whole development process, the interactions between developers and the governments are frequent. The governments use various permits to regulate the behaviors of the developers. The control index such as plot ratio, green ratio, building density, and building height restriction could not be exceeded. In addition, the negotiation space for developers has been decreasing these years, especially in terms of provision of infrastructures and public facilities. (Official B, 04-12-2015)
>
> Although the developers tend to negotiate with the governments in terms of the planning control over the land use and public facilities, it is very difficult to make changes due to the appealing procedures, consuming much time. The appeal could only be submitted to the planning committee half a year, and it required that no one oppose to the appeal and should be displayed publicly. If the planning committee rejected the appeal, the above procedure will go over again. Time is important to developers due to the amount of loan interest. (Official A, 04-12-2015)

Such comments show that the governments' regulation on service facilities in housing development could guarantee a relatively good quality of the residential environment. However, the government has not been aware of the effects of the service facilities on residents' health, especial the chronic diseases and emotional well-being.

7.2.3.3 Institutional Causes of the Failure of Optimization in Housing Operation

It has been demonstrated that there is weak operative transaction, and residents have to adapt passively to the physical residential environment without participating in the housing management affairs. This section examines the institutional environment of housing management to find out the institutional obstacles to the optimization. The interactions among housing managers, residents' committee, and residents are examined to find the causes of the limited residents' participation.

The literature shows that two key actors become quite influential in the operation of gated housing estates: the property management company (housing manager) and homeowner associations (HOAs). For HOAs, it has arisen as a brand-new neighborhood organization, oriented to the protection of residents' private property rights and neighborhood common property rights [16]. Although the government initially encouraged residents in SCHEs to set up homeowners' associations to strengthen neighborhood development, different interests across homeowners, developers and governments have resulted in clashes that have forced the government to change its attitudes [20]. In this study, no HOAs have been formulated in the three SCHEs, indicating the underdevelopment of civil society in the suburban housing estates. This may be one of the obstacles to residents' participation in housing management, indicating the limited channel for residents' engagement.

When comes to the conflicts between the housing managers, there is no institutional support for residents to play a significant role, discouraging residents' participation in housing management affairs. The barriers between housing managers and residents in housing management are hard to break without institutional innovations. For example, most conflicts between residents and housing managers are caused by the discrepancy between residents' expectation on the housing management and what the housing managers can afford to provide based on management fees, which is controlled by the government. The charging standard of management fees are stipulated in the property service contract at the time of sale, and residents do not want to adjust it. However, housing managers prefer to adjust the management fees according to the market conditions to make profits. Thus, it will cause dissatisfaction among residents if the housing managers try to raise the management fees, and the quality of housing management will deteriorate gradually if the management fee stays at the same standard over the years. For adjusting property fee, "Guangzhou property management ordinance" (2014) indicates that it could only be raised with the agreement of more than 50% of homeowners and the agreement of the homeowners who occupy more than 50% building area at the same time. This is very difficult to achieve, especially without the help of HOAs. As a result, the low management fee has negative effects on housing management, resulting in the dissatisfaction of the residents. The three SCHEs in this study are facing the declining quality of sanitation and maintenance. LNT is a representative case: the property fee in LNT has not been raised for more than ten years, leading to the continuously decreasing maintenance quality. The housing manager of LNT expressed his helpless:

> The service delivered by us is correlated with the property fee obviously. With limited property fee, the variety and quality of services that could be delivered are constrained. We need money to maintain the facilities and disposing of trash. However, we could only increase the property fee from the original price unless more than two-third residents agree with the incremental expenses, which is almost impossible. So we can only do our best to maintain the basic quality of the living environment. Although I know that there are many complains about the maintenance of green space and facilities in common area, we only focus on the most serious problem and we cannot afford to improve and make repairs to all of issues. (Housing manager of LNT, 27-11-2015)

Based on the above discussion, it is argued that the current institutional design of the property management negatively affects the interactions between residents and housing managers, generating conflicts and low participation of residents. Putting up with the conflicts and the limited ways of improving the situation, the willingness of residents to participate is reduced. In this context, improvements of residential environment through housing management are confined, making it hard to achieve the optimization of people-environment congruity.

Furthermore, the role of residents' committee and its relations with housing managers and residents are also found, implying to be adverse institutional environment in fostering residents' self-governing organization. As an executive branch of the local government, residents' committee usually cooperates with housing managers in housing management, especially when there are disagreements between residents and housing managers. Thus, there is cooperation between housing managers and the residents' committee in housing management. One officer from the residents' committee of LNT and one staff member from the property management company of GV offered comments:

> We interact with the housing manager frequently. Sometimes they help us to conduct some community activities, and we often help them to reconcile disputes. For example, we usually have to intervene in the conflicts stemming from the property fee. (Officer, LNT residents' committee, 27-11-2015)
>
> Residential committee is responsible for solving the neighborhood disputes, and all conflicts in the neighborhood should be settled by the residents' committee. The relationship between housing manager and residential committee is complementary, and the two organizations collaborate with each other in many matters pertaining to housing operation. (GV housing manager, 26-11-2015)

Such opinions suggest that housing managers and residents' committee share responsibility for housing management. Some scholars note that the state-managed residents' committees sometimes side with the home-owners and sometimes with the housing manager [11]. In terms of the attitude to the HOAs, the residents' committee could take over the functions of HOAs, which is not in favor of nurturing the HOAs. More fundamentally, the governments worry that a homeowners' group may overstep its authority by claiming to represent the residents, and they do not want to boost the capacity of homeowners to exercise power over their housing complexes [11]. From the perspective of the government, there is no reason to encourage the residents' self-organizations, and every reason to oppose them [11]. So a civil relationship of equality between the residents and housing managers has not been established [11]. It could be concluded that there is no institutional support for the growth of homeowners' associations, which makes it difficult for the residents to participate in housing management affairs, going against the optimization of residents-environment transactions.

Thus, the institutional environment of housing management, profit-driven housing manager and the control of residents' committee leave no space and way for the residents to form self-organizations and to participate in housing management affairs. This affects the people-environment congruity, reducing the opportunity to create a healthy neighborhood.

7.3 Conclusion

This chapter explains two crucial issues for the creation of a health-promoting environment: residents-environment transactions and institutional factors formulating the residential environment. Based on the examination of residents-environment transactions in terms of interpretive, responsive and operative modes, this chapter finds that residents of SCHEs have to passively adapt to the residential environment. Individual socioeconomic characteristics, experiences and demands are important for interpretative transactions, indicating the match degree between residents' cognition and expectation, which could in turn affect their health. Responsive transactions in SCHEs suggest that the physical activity of residents is an outcome of the residential environment in relation to constraints and opportunities. In addition, two reasons for the absent of operative transactions are pointed out: the limited channel for participating in housing management affairs, and the lack of initiatives of residents to engage in. On the whole, this chapter ascertains that the residential environment of SCHEs has shaped the way of life in particular the physical activity, affecting residents' health ultimately. In addition, psychological stress of SCHEs living is likely to be determined by the interpretive transaction, and the lack of residents' participation relieves the absent of civil society and the relatively satisfied housing management at the same time. Thus, the moderate effects of the passive adaption on residents' health are identified, which are the underlying obstacles to a healthier neighborhood.

In addition, this chapter examines the interactions across different stakeholders, based on which the institutional factors influencing residential environment are proposed. The institutional environment of housing management is not appropriate for a health-promoting environment. Contextually, institutional uncertainty and neoliberal policies for housing development have placed more emphasis on private profits rather than on human health. The default of local government and excessive dependence on developers for service facilities and the absence of civil society could be the main obstacles to a health-promoting environment. Notably, the institutional design of the provision of service facilities in Guangzhou (developer construct, transferring to the government) is a key reason for the dissatisfactory service facilities in SCHEs, especially for public transportation, fitness, and educational and daily commercial facilities. Governments' regulation could guarantee the basic standard of the quality of the residential environment, but without health-promoting elements especially those reduce chronic diseases and benefit mental health. What's more, the institutional environment for housing management could not promote the optimization of resident-environment transactions, with residents' committee and housing manager taking charge of all operation affairs and thwarting the establishment of homeowners' organizations. Thus, this study argues that there is weak institutional support for creating health-promoting environments in housing development and housing management.

References

1. Banerjee T, Baer W (1987) Variable constructs and images of the residential environment: Findings from Los Angeles. Percep Cognitive Image City, 271–303
2. Ekers M, Hamel P, Keil R (2012) Governing suburbia: modalities and mechanisms of suburban governance. Reg Stud 46(3):405–422. https://doi.org/10.1080/00343404.2012.658036
3. Evans GW (1987) Environmental cognition. Concept Publishing Company
4. Halpern D (1995) Mental health and the built environment: more than bricks and mortar? Taylor & Francis US
5. Humpel N, Owen N, Leslie E (2002) Environmental factors associated with adults' participation in physical activity: a review. Am J Prev Med 22(3):188–199. https://doi.org/10.1016/S0749-3797(01)00426-3
6. Lin F (2008) The study of countermeasure about the gread residential district in Guangzhou "Huanan Project." [Master thesis]. Tongji University
7. Meng M (2011) Public facilities comparison research on typical residential district for 'Huanan Plate' in Guangzhou [Master thesis]. South China University of China
8. Nasar JL (2000) The evaluative image of places. In: Person–environment psychology: new directions and perspectives, 2nd edn. Lawrence Erlbaum Associates Publishers, pp 117–168
9. Navarro V (2007) Neoliberalism, globalization, and inequalities. Routledge
10. Owen N, Leslie E, Salmon J, Fotheringham MJ (2000) Environmental determinants of physical activity and sedentary behavior. Exerc Sport Sci Rev 28(4):153–158
11. Read BL (2003) Democratizing the neighbourhood? new private housing and home-owner self-organization in urban China. China J 49:31–59. https://doi.org/10.2307/3182194
12. Shen J, Wu F (2012) The development of master-planned communities in Chinese suburbs: a case study of Shanghai's Thames town. Urban Geogr 33(2):183–203. https://doi.org/10.2747/0272-3638.33.2.183
13. Shen J, Wu F (2013) Moving to the suburbs: demand-side driving forces of suburban growth in China. Environ Plann A: Econ Space 45(8):1823–1844. https://doi.org/10.1068/a45565
14. Stokols D, Altman I (1987) Environmental psychology. Percep Cognitive Image City 12:23–90
15. Stokols D, Grzywacz JG, McMahan S, Phillips K (2003) Increasing the health promotive capacity of human environments. Am J Health Promot 18(1):4–13. https://doi.org/10.4278/0890-1171-18.1.4
16. Wang F, Yin H, Zhou Z (2012) The adoption of bottom-up governance in China's homeowner associations. Manag Organ Rev 8(3):559–583. https://doi.org/10.1111/j.1740-8784.2011.00277.x
17. Wang G, Liu S (2009) Guangzhou new town planning: the context of 'Asian Game' movement. City Plann Rev, 46–51
18. Wang YP, Shao L, Murie A, Cheng J (2012) The maturation of the neo-liberal housing market in urban China. Hous Stud 27(3):343–359. https://doi.org/10.1080/02673037.2012.651106
19. Wu F (2010) Gated and packaged suburbia: packaging and branding Chinese suburban residential development. Cities 27(5):385–396. https://doi.org/10.1016/j.cities.2010.06.003
20. Wu F, Shen J (2015) Suburban development and governance in China. In: Suburban governance: a global view. University of Toronto Press
21. Wu F, Zhang J (2007) Planning the competitive city-region: the emergence of strategic development plan in China. Urban Affairs Rev 42(5):714–740. https://doi.org/10.1177/1078087406298119
22. Yuan Q, Wei C (2011) From 'large community' to 'new town': reorganization of 'South China residential block.' J Urban Reg Plan 2:101–118
23. Zhang X (2000) Housing reform and the new governance of housing in urban China. Int J Public Sect Manag 13(6):519–525. https://doi.org/10.1108/09513550010356575
24. Zhang X (2002) Governing housing in China: state, market and work units. J Housing Built Environ 17(1):7–20

25. Zhang X (2006) Institutional transformation and marketisation: the changing patterns of housing investment in urban China. Habitat Int 30(2):327–341. https://doi.org/10.1016/j.habitatint.2004.02.008
26. Zhu J (2013) Governance over land development during rapid urbanization under institutional uncertainty, with reference to periurbanization in Guangzhou metropolitan region, China. Eviron Plann C Gov Policy 31(2):257–275. https://doi.org/10.1068/c11309

Chapter 8
Conclusion

Health implications of urbanization and urban environment have been popular multidisciplinary themes in academic literature. In China, suburbanization stimulated by the spread of commodity housing is distinctive, so its effects on human health are worthy of investigation to fill the research gap. Three commodity housing estates in the suburban Panyu District, Guangzhou were investigated and assessed in terms of their residents' health performance, its relationship with the characteristics of the residential environment, and residential behavior patterns, identifying the factors that affect health in relation to the residential environment and health-related residential behaviors. This study then interprets the factors that are crucial to the creation of health-promoting environment: examining the transactions between environmental and behavioral factors from a human ecological perspective, and explaining the institutional determinants of the residential environmental factors that affect health in housing development.

This chapter summarizes the major findings answering the three research questions. Based on the institutional factors shaping the residential environment, this chapter proposes the policy implications and recommendations for creating health-promoting environments. Finally, the empirical and theoretical contributions of this study to the literature are presented, followed by the limitations and suggestions for future studies.

8.1 Summary of Major Findings

This study answers three research questions about the effects of suburban living on residents' health in Guangzhou and about how the factors influencing health were formulated. This section summarizes the significant findings.

T. Zhang, *Towards Healthy Settlements*, Urban Sustainability,
https://doi.org/10.1007/978-981-97-1207-6_8

8.1.1 *Health Performance of Suburban Residents*

The first step is to evaluate the difference in health performance of residents in SCHEs of Guangzhou and explore the relationship between physical health and mental health.

8.1.1.1 Physical Health

Two popular variables are selected to indicate physical health of residents in this study: the incidence of obesity and self-reported general physical health. The results showed that the physical health of the residents was barely satisfactory. Frist, the prevalence of obesity was 15.3%, indicating a moderate prevalence of obesity compared with the Western situation. Second, most residents (71.7%) reported that they were in good physical health, 21.9% reported that they accomplished less than they would like and 13.3% said that they were limited in the kind of work or other activities due to the physical health issues. Only 2.9% of the residents claimed to be in poor physical health. For health residents, older residents were more likely to report better physical health than younger ones, possibly because older people were under less stress, indicating the effects of mental health on physical health.

Across the three cases, no significant differences were found in incidence of obesity and self-reported general physical health, implying that the differences in physical residential environment of SCHEs have no influence on the physical health. In addition, interviewees indicated that physical health lay on residents' mood, diet, and habit, in addition to the quality of the living environment.

8.1.1.2 Mental Health

Based on the literature, the assessment of mental health of residents focused on the prevalence of depression and the risks of mental disorders (RMD). This study found that the mental disorders and the risks for them in urban environment were underestimated.

The incidence of depression was 12.1%, much higher than other studies conducted in China with different measurements. For RMD, the percentages of residents reported feeling relative anxiety, stress, and claiming negative impacts of anxiety and stress on daily living were 7.9%, 16.0% and 7.0%, respectively. On the whole, the mean of RMD score as 3.98 suggested that the risks of mental disorders for residents in SCHEs were relatively low (on a scale of 5). Across the three SCHEs, significant differences in depression and RMD were found. The incidence of depression in AG Estate was much higher than that in LNT Estate and GV Estate; the RMD score of GV Estate was lower than that of the other two SCHEs.

8.1.1.3 Interaction Between Physical and Mental Health

The positive correlations between mental health and physical health were proved by this study. With the consideration of interrelations between physical and mental health conditions, statistical analyses suggested that depression and RMD were correlated with self-reported general physical health statistical significantly. In general, this study proves that positive moods, emotions and perceptions are of great importance to good physical health, and vice versa.

8.1.2 Factors Influencing Health

The second step of analysis addressed the research questions: *what and how have the factors stemming from suburban living affected the health performance of residents?* The factors influencing health were explored in terms of residential environment and health-related residential behaviors.

8.1.2.1 Influential Factors for Physical Health

Constrained by the cross-sectional research design, this study is incapable to provide the consequential evidence to ascertain direct effects of the residential environment on residents' physical health. Observations show that the three SCHEs are different in terms of the pedestrian-vehicles relation, street density and morphology, public transportation system, quantity and distribution of green space, quality of schools and daily commercial services, the sense of belonging and aesthetic style. However, no effect of these differences on residents' self-reported general physical health has been identified in this study. So this study argues that these differences may either have similar effects or have no effect on the general physical health, which needs to be differentiated in further studies.

Then this study focused on the influences of physical activity on the prevalence of obesity and self-reported general physical health. First, only walking time per weekday was found to be significantly associated with the prevalence of obesity. Although it is still debatable in terms of the casual direction between these two variables, this study found that 30 min of walking per day was associated with lower risks of obesity, confirming the argument that physical activity could reduce obesity risk. Age and gender were robustly correlated control factors. People between the ages of 50–59 had much higher possibility of being obese than did people aged 20–29. Men were at much higher prevalence of obesity than women, due to the differences in self-perception of their weight, in consciousness of or motives for weight control, and in stress and time for physical activity.

Second, two factors were found associated with the self-reported general physical health: driving time per weekday and the preference for physical leisure activity. Driving more than one hour per weekday and preferring to do physical leisure activity were associated with higher possibility of reporting good physical health. It is argued that driving may increase the self-actualization and provide a comfortable environment for long-distance commuting, reducing the psychological stress and improving the quality of life. In addition, the effects of the prevalence for physical leisure activity could be attributed to the benefits of physical activity on physical health, which has been confirmed by many studies.

These findings suggested that physical activity patterns of residents in SCHEs had few negative effects on their physical health, raising the question of whether residents of SCHEs were sedentary or car-dependent. So physical activity pattern was examined in terms of travel activity and physical leisure activity. For travel mode, it is found that 75% of residents used active travel modes (walking and cycling) and public transportation to get to work, suggesting they were not sedentary in travelling. For travel time, 46.6% of residents spent more than two hours travelling per weekday, while only 8.5% drive more than two hours per weekday, indicating that they are not car-dependent generally. For the physical leisure activity pattern, 68.6% of residents engaged in at least one type of physical leisure activity, among whom 56.7% participated in that activity for more than three times per week and 74.9% lasted more than 30 min. For the attitudes towards physical leisure activity, 73.2% of the residents expressed a preference, and 50.2% reported that they conducted the activities regularly. Based on these findings, this study argues that residents of SCHEs are neither sedentary nor car-dependent. By the same token, it is not a health-promoting lifestyle since commuting may exert a health toll and the intensity and frequency of the physical activity may not be adequate for keeping good health.

8.1.2.2 Influential Factors for Mental Health

The effects of suburban living on the prevalence of depression and RMD were examined in relation to residential environment and health-related residential factors. First, three factors were significantly associated with the prevalence of depression: neighborhood type, time spent on public transportation, and the variety of physical leisure activity. Living in AG Estate, travelling by public transportation more than 30 min per day, and engaging in no physical leisure activity were associated with greater likelihood of being depressed. For the effects of physical residential environment, the association between neighborhood type and depression is the result of selection but not causation, because AG Estate had better quality of the residential environment, especially in terms of fitness facilities, green space and landscape. For the effects of physical activity, the causal relationship between time spent on public transportation and depression could be explained by this study. It is argued that poor environmental quality of public vehicles and uncontrollable transfer time may increase travelers' psychological stress, exacerbating the symptoms of depression. At the same time, it is possible that the depressed people have limited choices in travel mode but had to rely on public transportation, indicating the selection mechanism of the association. In addition, although the causal relationship between variety of physical leisure activity and depression could not be ascertained by this study, it is clear that physical leisure activity could mitigate the prevalence of depression based on other studies and the qualitative data from interviews.

Furthermore, two groups of environmental and behavioral factors were found associated with RMD at statistically significant level. Eight factors were found in the residential environment: neighborhood type, housing type, perceived accessibility of designations, the place chosen for physical leisure activity, satisfaction on educational and daily commercial facilities, green space, and the sense of belonging. Four factors were identified in residential behaviors: walking time per weekday, total travel time per weekday, regularity of physical leisure activity, and variety of neighbors' interactions.

For the effects of physical residential environment, living in AG and in a villa or townhouse was associated with lower risks of mental disorders than living in LNT and in a high-rise environment. The causation could be inferred from the literature which claims that a better living environment and social status as represented by the housing estates could decrease stress and improve the sense of identity and control over individual's life, thus benefiting mental health.

For the effects of perceived residential environment, this study proposes the following arguments. Good accessibility of the designations and convenient and attractive places for physical leisure activity could promote walking activity and increase the possibility and frequency of physical leisure activity, thus reducing risks of mental disorders. Diverse and available educational and daily commercial facilities could relieve social and economic pressures, decreasing risks of mental disorders. More green space could relieve psychological stress and increase physical activity and social interactions, leading to the reduction of risks of mental disorders. The sense of belonging could eliminate loneness, and build social support and benefit neighborhood cohesion, thus reducing the risks of mental disorders.

For the effects of physical activity, walking more than 30 min per weekday and taking regular leisure activity were associated with lower risks of mental disorders, which could be explained by the positive effects of physical activity. In addition, travelling more than two hours per weekday was associated with higher risks of mental disorders. Excessive travel time could increase risks of mental disorders, because it may increase stress from the uncomfortable and tiresome journey and it could reduce the time spent on physical activity and social interactions. Furthermore, participating in three types or more informal interactions with neighbors was associated with lower risks of mental disorders. This association could be explained by the literature and interviews: informal interactions may increase the perception of neighborhood cohesion, cultivate a neighborhood network and promote social support, which could contribute to better mental health.

8.2 Summary of the Factors Influencing Health

According to the connotation of the identified factors influencing health, six grouping factors are aggregated in Table 8.1: neighborhood and housing types, public service facilities, design features, travel time, physical leisure activity, and social interactions. With regards to the frequency of the appearance of each factor, the weight of each is calculated. It is argued that travel time of different travel modes is the most influential factors in residents' health, followed by the public service facilities, indicating the significance of physical activity to residents' health.

For the associations between these factors and health performance, the causation mechanisms are discussed with the help of interviews and the literature. Five mediators between the environmental and behavioral factors and health outcomes are proposed: effects of daily routine, benefits of physical activity, effects of psychosocial stress and life stressor, influence of identity and symbolic meaning and self-esteem, and impact of social support and neighborhood cohesion (Fig. 6.17). Specifically, travel time, pattern of physical leisure activity, and design features have influences on the benefits of physical activity gained by residents, which in turn affect residents' health. In addition, travel time, public service facilities, and design features could affect the residents' psychosocial stress, influencing the mental health in particular. Design features and social interactions influence the social support that residents could get and the form of neighborhood cohesion, affecting mental health as well. Taken together, it is argued that benefits of physical activity and effects of psychological stresses are the two most influential explanations between suburban living and

Table 8.1 The aggregated factors influencing health

Aggregation factors	Influential factors of the four health indicators	Weight
Neighborhood and housing type	Neighborhood type Housing type	3/18
Public service facilities & design feature	Accessibility of destinations Place for leisure activity Education facilities Commercial facilities	4/18
Design feature	Quantity of green space Sense of belonging	2/18
Travel time of different travel modes	Walking time per day Driving time per day Time on public transportation Overall travel time	5/18
Pattern of physical leisure activity	Preference for physical leisure activity Variety of physical leisure activity Regularity of physical leisure activity	3/18
Social interactions	Variety of informal neighbors' interaction	1/18

health. Thus, the suburban living environment and resultant suburban lifestyle are ascertained to be associated with residents' health, especially influenced or mediated by the quality of public service facilities, design features, travel activity and physical leisure activity. In line with extant studies, the sensible built environment of suburban neighborhood could scale back negative health effects through providing enough sports facilities and more green space [7]. In this sense, compared with living environment in urban core areas, the suburban neighborhoods have more advantages in terms of the opportunities for physical leisure activity, greenery space and aesthetic landscape, and the sense of belonging, which mitigate for the disadvantage of automobile dependence and contribute to the residents' health.

8.2.1 Residents-Environment Transactions

Considering the interactions between environmental factors and residential behavioral factors, the current modes of residents-environment transactions were examined to ascertain the distance from SCHEs towards health-promoting environments.

First, based on residents' cognitions on public service facilities and design features, this study argues that the interpretive transaction is mainly affected by socioeconomic characteristics of residents (especially age, homeownership, and hukou status), individual's experience and demands and sociocultural understanding. The coherence of the physical residential environment and aesthetic visual characteristics are influential in the interpretive transaction between residents' affective appraisal and physical residential environment.

Second, for the responsive transaction, physical activity of suburban residents is affected by physical and perceived residential environment, and residents have to adapt to the residential environment passively. In terms of travel activity, residents passively adapted to the physical travel environment in terms of the selection of travel mode. Three characteristics of transportation environment, comfort, convenience, and flexibility, are determinative to residents' travel choice. Residents choose to adjust their travel demands to adapt to the fixed transportation environment. In addition, the pattern of physical leisure activity is a result of passive responsive transaction. Residential environment could affect its regularity and variety easily, so residents adjust their behaviors according to both constraints and opportunities for physical leisure activity in the residential environment.

Third, this study finds no operative transaction between residents and the physical residential environment. Two possible reasons are proposed: residents have inadequate motivation to participate in housing management affairs because there were no big issues and rigid conflicts in housing management process; and there are no effective ways for residents to engage in with the underdeveloped civil society [6].

Therefore, residents-environment transaction mode in SCHEs is a passive adaption one, indicating the dominance of the residential environment in forming residents' cognition and behaviors. This chapter ascertains that the residential environment of SCHEs has shaped the way of life in particular the physical activity,

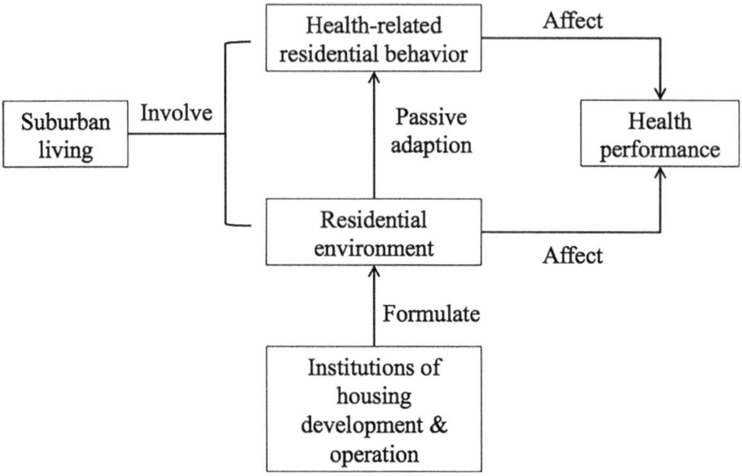

Fig. 8.1 Relations between the key components of the findings

affecting residents' health ultimately. With the relatively satisfied housing management, the moderate effects of the passive adaption on residents' health are identified. The passive adaption process between residents and their environment is an obstacle to achieve people-environment congruity to create a health-promoting environment, though with only moderate effects on residents' health and well-being.

On the whole, Fig. 8.1 illustrates the one-way relations between environmental and residential behavioral factors. The residential environment has both direct and indirect influences on residents' health. Public service facilities and design features are of particular significance. In this context, this study points out that the institutional factors formulating the physical residential environment were the underlying reasons for the partial failure to produce a health-promoting environment (Fig. 8.1).

8.2.2 Institutional Factors Affecting the Residential Environment

The third step of analysis concerns the research questions: *How have the institutional factors within housing development created the neighborhood living environment?* The institutional factors in suburban housing development were examined in terms of the initiatives of developing SCHEs, the decision-making of the SCHEs and housing management. Roles of local government, developers, planners, housing managers, and residents' committee and ordinary residents were interpreted. This study argues that the institutional environment for commodity housing development is not conducive to the creation of a health-promoting environment.

The role of the state is influential in the initiatives of suburban commodity housing development, which has been under a transition with the establishment of neoliberalism globally. With neoliberal urban development strategy in China, dual initiatives of governments to develop suburban commodity housing development are identified: state entrepreneurialism and growth machine. This leads to the transitions of state-market relations and changes of power relations within government hierarchy, thus creating institutional uncertainty in a short transition period. In addition, the neoliberalism-oriented strategy in the housing market emphasized private profit rather than public interests such as human health.

Stimulated by consumer sovereign, developers in the decision-making of SCHEs are dominant, determining the physical residential environment. Case studies suggested that they based their choice of the type of housing products on market conditions, and the demands of potential homebuyers determined the characteristics of physical residential environment greatly. In terms of the provision of public service facilities, the institutional design is determinative: "developers construct, transferring to the government". The governments' regulation could guarantee the basic public service facilities, but the excessive dependence on developers in construction could lead to the problems in service quality and service type. There are two scenarios, for the facilities transferred to governments free of charge, developers have no impetus to build quality ones; for the facilities hold by developers, developers may make efforts to make profits from them by attracting homebuyers.

The institutional environment in housing management failed to promote optimization transaction between residents and environment. Housing managers and residents' committee are in charge of housing management, and there was limited channel and no motivates but various constraints for ordinary residents to participate in. This institutional environment could not cultivate residents' self-organizations, which are crucial to residents' participation. Thus, this study argues that institutional regulations in housing management has blocked the operative transaction between ordinary residents and residential environment, which is adverse to the creation of health-promoting environments.

These findings have demonstrated the propositions that were raised in Chap. 4. First, under the neoliberal urban strategy, the initiatives of the local government developing SCHEs were stimulated by the desire of promoting economic growth or gaining resources for regional and global competition. For the second and third propositions, current institutional design for the SCHEs development empowered developers to make decisions on both the types of housing products and public service facilities, both of which were affected by the market. The governments only intervened with regulations to ensure the basic quality of the living environment. Planners were in an affiliated position, taking orders from both developers and governments. As a result, the health needs that were consistent with the profits of developers could be easily satisfied, such as attractive green landscapes that could attract homebuyers. Health needs that were inconsistent with the profit making were hardly to be met such as fitness facilities and quality schools. In addition, with housing manager and residents' committee dominating the housing management and without the effective participation of residents, the operative transactions between residents and environment are hard to be created, which has negative effects on the creation of health-promoting environments.

8.3 Policy Implications and Recommendations

Drawing on the findings, this study identifies several policy implications and recommendations for achieving a healthy neighborhood living environment for middle-income residents suburban China. With the passive adaptation transaction between residents and their neighborhood environment, the physical characteristics of the residential environment are considered crucial to health. So the factors influencing health identified in this study are vulnerable to the institutional factors that formulate the physical environment. Policy recommendations focus on institutional and practical aspects, such as an innovation institutional environment for creating health-promoting environments and the implementation of healthy planning and design. The institutional environment of urban development should embed a health dimension. As pointed out by @@[4], the creation of healthy living environment requires a new perspective of health and well-being in government. Human health should be the central goal of the governments and the whole society, since the important health determinants lie on the sectors other than health [4]. "All sectors and levels of government and society should recognize health as both a means to reach their own goals but also as a responsibility towards the whole society" [4], p. 82). Thus, health could not be achieved by a single sector alone, indicating that it is necessary to enlist participation and collaboration between the public and private sectors, and between health and non-health sectors [4]. The institutional innovation in the structures and mechanisms for creating health-promoting environments is thus necessary. In housing development, human health should be regarded as an overriding public interest. Key stakeholders in the housing development should recognize the complexity of human health and its determinants [4].

First, given that the state is capable of promoting rapid and widespread change in a society in China [3], governments should take more responsibility for meeting health needs through regulations on land use, building standards, transportation, and public service provision, all of which can protect human health. In addition, governments should intervene in suburban commodity housing development and especially in the provision of public service facilities. The health-promoting environments could be created taking advantages of the governments' interventions that prevent market failures.

Second, considering the commodity feature of the SCHEs and the profit-driven nature of developers who focus on short-term economic advantage over long-term sustainability [5], an indirect but effective way for governments intervening in the decision-making of developers is to educate homebuyers. Education could awake the health consciousness of the public. In particular, the middle-income population could be educated on a health-promoting environment and the many determinants of good health, which could influence their choice of a home and guide the market to provide healthier residential environment. For example, governments could help to reduce obesity by emphasizing the importance of physical activity [3].

Third, the civil society's role deserves more attention. Planners and residents should be empowered to build a civil society, with the backing of the governments rather than to suppress. Many challenges for health could be better resolved by the involvement of civil society and private sector [4]. So public engagement has an important role in the formation of people-environment congruity. In particular, residents should have a greater role in housing management, establishing the operative transactions between residents and the residential environment. This recommendation could see its root in the governance for health, proposing to rebuild institutions to increase stakeholders' participation and to cultivate a network government for the creation of a healthy urban environment. For governments, it could be beneficial to conduct the institutional innovation, since working with the communities and with wider distribution of power in society, the governments could enhance its legitimacy, generate additional social capital and gain valuable resources (Kickbusch and Gleicher [4]). Network governance for health could help to promote the optimization of the residents-environment transactions, thus benefiting the creation of health-promoting environments.

In practical terms, greater efforts should be devoted to invigorate the collaboration between urban planning and public health to guide suburban housing development. A healthy cities approach should be operationalized in planning and design, by emphasizing active living, especially in the creation of buildings, streets, land use patterns, active travel modes, opportunities for physical leisure activity, and quality educational and daily commercial facilities [2]. Clear operational criteria of healthy planning and design are needed in the creation of healthy neighborhoods. Since developers prefer to apply clear and simple economic criteria to the development, the implicit criteria of health benefits should be avoided [5]. The operation criteria for health-promoting environment should be elaborated in detail, avoiding the disparities between understanding and operationalizing. It is recommended that planners and governments should go beyond the symbolic use of healthy neighborhoods to establish comprehensive and practical development guidance strategies.

8.4 Contributions to the Literature

This study contributes both empirically and theoretically to literature. Overall, this study complements the limited literature on health effects of the dynamic urban spatial expansion in developing countries in Asia (China), which are distinctive from the existing findings in western literature. Focusing on urbanization and suburbanization and their resultant implications, health effects of the suburban living of the new middle-class in Guangzhou were examined to reveal the characteristics of the suburban lifestyle and its impacts on residents' health.

First, this study clarifies that the definition of suburbs and suburbanization should be based on the locality, which makes the essence of the suburbanization in China different from that of western contexts. It is argued that Chinese suburbanization could be seen as decentralized urbanization, and the suburbia is an outcome of urbanization. Territorial and social attributes of suburbs are highlighted in China, especially with the boom of suburban commodity housing development after the economic and land reforms.

Second, this study points out that the physical health performance of suburban residents was barely satisfied, which is different from that of western population, especially from the population of North America in terms of the prevalence of obesity. Accordingly, it is pointed out that the physical activity pattern of suburban residents in China was more active than that of suburban residents in North America, particularly in terms of travel activity.

Third, this study produces significant knowledge on the suburban living in Guangzhou in relation to obesity, self-reported general physical health, depression, and risks of mental disorders. Distinctive health determinants associated with physical and mental health performance were unfolded, which has been less discussed in western literature. It is found that public service facilities and time spent on different travel modes are the most influential factors for the middle-income residents in suburban neighborhoods. Additionally, this study derives several different findings from the existing literature, such as the positive effects of driving on general physical health, and that residents in SCHEs are not sedentary or car-dependent. These findings may rectify some prejudice and verify the characteristics of the suburban lifestyle in China empirically, based on which the conceptual understanding of the health determinants could be expanded.

Forth, this study identifies the interactions between residents and their residential environment from a human ecological perspective, which expands the application of people-environment transaction modes at neighborhood level. Few studies have tried to account for people-environment transactions and their health effects. This study finds that the SCHEs residents have passively adapted to their environment without resorting to the operative transaction. This finding ascertains the indirect effects of residential environment on residents' health by affecting the residents' behaviors.

Finally, this study also fills the research gap of the limited explanation on institutional factors for health-promoting environments in housing development. Through examining the institutional issues in initiatives and decision-making for suburban housing development and housing management, the roles of and interactions between the key players in suburban housing development are revealed. This helps to better understand the institutional obstructions for creating healthy neighborhoods, based on which policy recommendations are proposed in terms of institutional innovations. This may contribute to the building or rebuilding of network governance, thus benefiting the new commodity housing development in future in creating health-promoting environments.

To summarize, compared with western literature, this study fills up urban health research in China, showing profound significance in finding the health determinants of urban development in the context of rapid urbanization in Asian developing countries both empirically and conceptually.

8.5 Limitations of This Study and Future Suggestion

Despite the significant findings and contributions to the literature, this study has several unavoidable limitations. The most obvious limitations are methodological. Since this is a cross-sectional study, it is hazardous to draw conclusions about causality. This is inevitable since the longitudinal studies are very scarce and have a high requirement for data collection, especially focusing on the relationship between housing environment and mental health [1]. So the arguments made on the causation mechanism in this study are mainly drawn from the interviews and the supports stemming from literature.

In addition, there are several limitations stem from the data collection, especially in terms of the questionnaire survey. Constrained with time, available resources, and the gated nature of the SCHEs, this study resorted to convenience sampling, which is not random sampling. In addition, relatively small sample size was set and only 1% of residents in the three neighborhoods were under surveyed. Although it has been justified with the low level of expected variation of the data due to the relative homogenous socio-economic characteristics of neighborhoods, the limited sample size and non-random sampling may still harm the accuracy and reliability of the findings and bias the generalization of the results.

Furthermore, the health measurements in this study are not perfect, especially that the measurement of RMD was developed based on the literature, without a validation by medical experts. Although it could be justified that RMD is not the same as the clinical diagnosis of mental disorders, the precision of the measurement could be unsatisfactory. Moreover, the validity of the questionnaire is affected by the privacy of mental disorders, leading to much missing data about depression, and it was difficult to interview residents who were depressed.

Another limitation lies on the research scope. This study centers on three SCHEs in Panyu district, and the historical links between Panyu and Guangzhou make the three cases distinctive, especially in terms of the roles and functions of the local governments. Thus, it may not be appropriate to generalize some of the findings of this study to the SCHEs elsewhere in China.

Based on the limitation and constraints of this study, some suggestions are made for the future urban health studies. Foremost, longitudinal studies should be conducted to determine the causal relationships between the urban environment and human health. Considering the limitation on the research scope, diverse cases in suburban areas could be studied. In addition, future studies should focus on other regions or metropolitan areas in China, which could shed light on the universal relations between the suburban living and human health, benefiting the generalization of the findings which may contribute to the theoretical foundation of urban health. Furthermore, collaborations between experts in urban studies and public health should be encouraged to strengthen the precision and validity of the measurements of health and environments. Interdisciplinary research in urban planning and public health in China should be enhanced in the interest of healthy urban development. Finally, based on the arguments that institutional environment of housing development is crucial to the creation of health-promoting environments, more attentions should be paid to the institutional innovations in housing development to create health-promoting environments.

Accordingly, the future research for postdoctoral work focusing on the urban environment and public health should be optimized in terms of adopting the longitudinal research design, redefining the research scope and highlighting the geographic disparity, employing random sampling, improving health and environment measurements, and underlying the causal relationships as much as possible. The locality, especially the sociocultural and institutional characteristics, should be treated as significant points in explaining the health determinants in Chinese urban development, which could be compared with that in western literature, contributing to the theoretical framework for urban health in developing countries to some extent.

8.6 Conclusion

This chapter summarizes the major findings of this study on residents' health in SCHEs, their relations to the residential environment and health-related residential behaviors, and institutional factors that shape the residential environment. Based on the institutional factors in suburban commodity housing development influencing the creation of health-promoting environments, policy implications and recommendations are then proposed. Afterwards, the study's contributions to the urban health literature are presented, followed by the discussion of the limitations and suggestions for the future studies.

References

1. Dalgard OS, Tambs K (1997) Urban environment and mental health: a longitudinal study. Br J Psychiatry 171(6):530–536. https://doi.org/10.1192/bjp.171.6.530
2. Day K, Alfonzo M, Chen Y, Guo Z, Lee KK (2013) Overweight, obesity, and inactivity and urban design in rapidly growing Chinese cities. Health Place 21:29–38. https://doi.org/10.1016/j.healthplace.2012.12.009
3. French P, Crabbe M (2010) Fat China: how expanding waistlines are changing a nation. Anthem Press
4. Kickbusch I, Gleicher DE (eds) (2012) Governance for health in the 21st century. WHO Regional Office for Europe
5. Seabrooke W, Yeung SCW, Ma FMF, Li Y (2004) Implementing sustainable urban development at the operational level (with special reference to Hong Kong and Guangzhou). Habitat Int 28(3):443–466. https://doi.org/10.1016/S0197-3975(03)00043-2
6. Wu F, Shen J (2015) Suburban development and governance in China. In: Suburban governance: a global view. University of Toronto Press
7. Zhang S, Li J, Wang L, Kwan MP, Chai Y, Du Y, Zhou K, Gu H, Sun W (2023) Examining the association between the built environment and active travel using GPS data: a study of a large residential area (Daju) in Shanghai. Health Place 79:102971

Appendix A

Luoxi New Town (LNT)

Promoted by the completion of Luoxi Bridge in 1988, Luoxi New Town has been planned and developed since1989 under the cooperation of Hong Kong developers and local developers, aiming to release the housing shortage and turn the Shaku Island to a new town of Guangzhou. Covering around 185 hm^2, LNT is comprised of several housing estates (Ruyi Garden, Rainbow Garden, Jixiang Estate, Luohu Estate, and Luotao Estate, etc.) and involving different housing types including villas, multiple-level residential buildings, and high-rise residential buildings. The whole LNT provide 17,271 housing units, housing 57,017 population,[1] and administrated by three residential committee. While the sampling area in this study takes more than one-third area of the LNT, with the number of housing units of 9,353, and housing for 22,230 residents (Table A1).

Table A1 Basic information of Luoxi New Town (LNT)

SCHEs	Housing units	Population	Site area (hm^2)	Floor area ratio	Building density	Green ratio (%)
LNT (sample area)	9,353	22,230	47.03	1.60	–	20

[1] *Source* http://luopu.panyu.gov.cn/.

T. Zhang, *Towards Healthy Settlements*, Urban Sustainability,
https://doi.org/10.1007/978-981-97-1207-6

龙轩美容美发
旧货

Agile Garden (AG)

Agile Garden had been developing by Agile Property Holdings Limited from 2000 to 2010. It is located in the Xingnan Road, Nancun town, with a mountain on its south side. The whole housing estate is comprised of 15 residential clusters, covering an area of around 314.9 hm^2. There are overall 11,230 housing units, housing around 29,520 residents.[2] To be clear, the size of resident population is bigger than the official statistics, which is as huge as around 42,000 according to the director of the residents' committee. The building density of overall Agile Garden is 13.80%, the floor area ratio of 1.36, and with the green ratio of 36.80%, indicating the relative lower density (Table A2). There are various types of housing in Agile Garden, including villas, townhouse, and high-rise buildings.

Table A2 Basic information of Agile garden estate

SCHEs	Housing units	Population	Site area (hm^2)	Floor area ratio	Building density (%)	Green ratio (%)
AG	11,230	29,520 (42,000)	314.9	1.36	13.80	36.80

[2] http://www.nancun.gov.cn/ShowInfo.aspx?id=20120704102500265554943000625000.

Asian Game Village (GV[3])

Asian Games Village (GV) was developed for the 16th Asian games in 2010. Located in the south of Town of Shilou, GV has been developed on hopes of expediting the new town development and population movement. The post-game utilization of the GV and the local development has been considered to large extent ever since the stages of site selection and plan making. Specifically, GV was planned and developed in a sustainable and eco-oriented way, such as promoting the using of bicycle and public transportation instead of private cars and adopting various green and environmental techniques. In addition, the provision of public facilities is another attractive condition for residents, including the glamour schools and hospital. As a result, GV is attractive to certain population in terms of good environmental quality, public facilities, and public transportation system.

GV covers 270 hm^2 and comprises of four housing estates: Media Village, Athletes Village, Technical Officials' Village and The HIGHEST (Fig. 3). For the physical characteristics, with the relative low floor ratio (1.66) and building density (12.2%), GV enjoys high-rise residential buildings and high green ratio (54.2%) (Table 3). There are overall 11,687 housing units in GV, however, based on the interview with

[3] It is abbreviated as GV to differentiate with the abbreviation of AG.

housing managers of the four estates, it is found that the occupancy rate is relatively low: the occupancy rate of Media Village is around 80%, while the occupancy rate of Athletes Village and Technical and officials Village is around 60%. For the HIGHEST, only around 200 households have checked in by May 2015 (Table 4). Thus, it could be estimated that around 5807 housing units are occupied in the whole neighborhood. Since there is no census data for the population size, based on the average number of persons per household,[4] the number of residents living in this neighborhood is estimated as around 17,421.

Table A3 Basic information of Asian game village

SCHEs	Housing units	Population	Site area (hm^2)	Floor area ratio	Building density (%)	Green ratio (%)
GV	11,678	17,421	270	1.66	12.2	54.2

Table A4 Information of residents in Asian game village

Residential cluster	Numbers of housing units	Occupancy rate
Media village	3800	80% (3040)
Athletes village	3598	60% (2159)
Technical and officials village	680	60% (408)
The highest	3600	Only around 200 households have checked in by May 2015
Total	11,678	5807 units

Sources Interview with the management company of GV

[4] Average number of persons per household is estimated based on the Sixth Census of Panyu District. It is around 3 persons per household in Panyu District (1,764,828/586,086).

Appendix B
Dummy Variables

See Tables B1 and B2.

Table B1 List of the perceived residential environment variables

Dimensions	Original variables	Dummy variables/measurement
Walking environment (transport environment)	1. Accessibility to service facilities *(Score ranging from 1 to 5, with higher score represents better accessibility)*	**Continuous variable** = (Accessibility to subway station + accessibility to bus stop + accessibility to retail and market + accessibility to leisure facilities + accessibility to kindergarten and primary school + accessibility to clinic or pharmacies + accessibility to community services)/7
	2. Perception on the quality of walking surface (*Statement: the walking surface is suitable for walking*)	Agree (0), Not sure, disagree

(continued)

T. Zhang, *Towards Healthy Settlements*, Urban Sustainability,
https://doi.org/10.1007/978-981-97-1207-6

Table B1 (continued)

Dimensions	Original variables	Dummy variables/measurement
Leisure environment	3. Place preference for leisure activity	Within neighborhood (0), Outsider neighborhood, Others
	4. Perception on the quantity of open space *(Statement: There are enough outdoor open space (square, walking path,* etc.*) to do leisure activity)*	Agree (0), Not sure, disagree
	5. Perception on the quantity of fitness facilities *(Statement: There are* ***not*** *enough fitness facilities within the neighborhood.)*	Agree (0), Not sure, disagree
	6. Satisfaction on the quality of space and facilities *(Statement: I am satisfied with the quality of the open space and fitness facilities for physical exercises)*	Agree (0), Not sure, disagree
Service environment	7. Places chosen for medical health issues	Community clinics (0), Hospital in Panyu district, Hospital in urban core area, Others

(continued)

Table B1 (continued)

Dimensions	Original variables	Dummy variables/measurement
	8. Place chosen for medical health issues	Community clinics (0), Hospital in Panyu, Hospital in urban core area, Others
	9. Satisfaction on health service facilities *(Statement: There are not enough health facilities, which could not satisfy my demand)*	Agree (0), Not sure, Disagree
	10. Satisfaction on daily commercial facilities *(Statement: There are not enough commercial facilities, which could not satisfy my demand)*	Agree (0), Not sure, Disagree
	11. Satisfaction on education facilities *(Statement: The schools within this neighborhood could **not** satisfy my demand.)*	Agree (0), Not sure, Disagree
Intangible environment	12. Perception on green space *(Statement: the green space is not enough in this neighborhood)*	Disagree (0), Not sure, Agree
	13. Perceived density *(Statement: I feel the population density is all right, neither too crowd or too deserted)*	Agree (0), Not sure, Disagree
	14. Perceived safety *(Statement: The neighborhood is safe from crime)*	Agree (0), Not sure, disagree

(continued)

Table B1 (continued)

Dimensions	Original variables	Dummy variables/measurement
	15. Perception of ambient environment *(Statement: The ambient environment makes me comfortable (good air quality, good acoustic environment, good hygiene condition)*	Agree (0), Not sure, disagree
	16. Perception on public realm *(Statement: There are enough social space within the neighborhood for neighbors' interactions and community activities.)*	Agree (0), Not sure, Disagree
	17. Sense of aesthetic *(Statement: The appearance of the landscape and architecture are characteristic and looks good to me.)*	Agree (0), Not sure, Disagree
	18. Sense of belonging *(Statement: The design of this neighborhood makes me have a feeling of attachment and belonging.)*	Agree (0), Not sure, Disagree

Table B2 List of health-related residential behavior variables

Dimensions	Original variables	Dummy variables/ measurement
Travel activity	1. Travel type for commuting	Active travel mode (0), Public transportation, Private car, Others or multiple choice
	2. Time spent on walking in one typical weekday	<30 min (0), ≥30 min
	3. Overall travel time (destinations to destinations)	<2 h (0), ≥2 h
	4. Willingness for walking in residential living *(Compared to other means, I prefer to walk to destinations in daily life.)*	Agree (0), Not sure, Disagree
Leisure activity	5. Variety of attended type of leisure activities	No leisure activity (0), one type, two types
	6. Frequency of doing leisure activities	Everyday (0), more than 3 times a week, once or twice a week, once two weeks, once a month or less
	7. Duration of doing leisure activities	Less than 30 min (0), 30–60 min, more than 60 min
	8. Preference for leisure activities *(Statement: I love doing leisure activities such as physical exercises)*	Agree (0), not sure, disagree
	9. Regularity of leisure activities *(Statement: I did physical exercises regularly during past six months)*	Agree (0), not sure, disagree
Neighbors' interaction	10. Variety of informal neighborhood interactions	One type and less (0), Two types, Three types and more
	11. Frequency of informal neighborhood interactions	Everyday (0), More than 3 times per week Once or twice per week, Once per month, Never
	12. Willingness of joining informal neighborhood interactions *(Statement: In my spare time, I'd rather to stay at home than to interact with my neighbors)*	Agree (0), not sure, disagree

(continued)

Table B2 (continued)

Dimensions	Original variables	Dummy variables/ measurement
	13. Variety of community activities	None (0), One type, Two types and more
	14. Frequency of attending community activities	Never (0), Once per month or less, More than once per month
	15. Willingness of attending community activities *(Statement: I have **no** interests in the community activities)*	Agree (0), not sure, disagree
	16. Overall neighborhood cohesion	Continuous variable
Participation in neighborhood operation	17. Variety of participation in neighborhood operation	None (0), None type, Two types and more
	18. Frequency of participating neighborhood operation	Never (0), Less than once per month, Once per month, More than once per month
	19. Concerns for neighborhood operation *(I don't care the neighborhood operation and related affairs at all)*	Agree (0), not sure, disagree
	20. Satisfaction with the neighborhood operation *(Statement: I am **not** satisfied with the housing management)*	Agree (0), not sure, disagree

Appendix C
Questionnaire Survey

1. **Questionnaire for residents**

Respondent should be residents of the neighborhood who live here longer than six months, and who are older than 20 years old (included).

Screening questions

1. **Age:_______**
2. **Gender: M/F**
3. **How long have you been living in this neighborhood**: _______months/years
4. **Your income per year (after tax):_________RMB/year**
5. **Education status**

   (1) Bellow primary school education
   (2) Primary school grade
   (3) Junior high school degree
   (4) High school degree
   (5) Junior college degree
   (6) Bachelor degree
   (7) Above bachelor degree

6. **Where do you work (If you do not have a job, select 1)?**

   (1) Don't have a job
   (2) Within the neighborhood
   (3) Panyu
   (4) Nansha
   (5) Huangpu
   (6) Tianhe
   (7) Haizhu
   (8) Liwan
   (9) Yuexiu

T. Zhang, *Towards Healthy Settlements*, Urban Sustainability,
https://doi.org/10.1007/978-981-97-1207-6

(10) Others, please specify:________

7. **What is your household composition?**

   (1) Singe family
   (2) One couple family
   (3) Core family (one couple and their unmarried children)
   (4) Three-generation family
   (5) Grandparents and their unmarried grandchildren
   (6) Others, specify_________________.

8. **What kind of housing types do you live in?**

   (1) Villa or townhouse
   (2) Multiple building (around 6 floors)
   (3) High-rise building (around 11 floors)
   (4) High-rise building (around 17 floors)
   (5) Super high-rise building (around 30 floors)

9. **Are you a tenant?**
   (1) Yes (2) No
10. **Do you have *hukou* registered in Guangzhou?**
   (1) Yes (2) No

Part 1: Health status of residents

1. **Your height:_____cm; weight_____kg; BMI:_______ Body fat rate:______**
2. **How do you feel your general physical health?**

   (1) Excellent
   (2) Very good
   (3) Good
   (4) Fair
   (5) Poor
   (6) Cannot judge

3. **During the *past 4 weeks*, have you had any of the following problems with your work or other regular daily activities?**

		Yes	No
As a result of your physical health	*Accomplished less* than you would like	1	0
	Were limited in the *kind* of work or other activities	1	0
As a result of any emotional problems	*Accomplished less* than you would like	1	0
	Were limited in the *kind* of work or other activities	1	0

4. **How much of the time during the past 4 weeks…**

	All of the time	Most of the time	A good bit of the time (Around half the time)	Some of the time (Less than half the time)	None of the time	Don't know
1. Have you felt calm and peaceful?	6	5	4	3	2	1
2. Have you felt anxious?	1	2	3	4	5	6
3. Did you feel stressful?	1	2	3	4	5	6
4. How much of the time do anxiety and stress interfere with your daily living?	1	2	3	4	5	6

5. **How often you have felt this way during *the past week*. (Radloff, 1977)**

	Less than 1 days	1–2 days	3–4 days	5–7 days
1. I was bothered by things that usually don't bother me	0	1	2	3
2. I did not feel like eating; my appetite was poor	0	1	2	3
3. I felt that I could not shake off the blues even with help from my family or friends	0	1	2	3
4. I felt that I was just as good as other people	0	1	2	3
5. I had trouble keeping my mind on what I was doing	0	1	2	3
6. I felt depressed	0	1	2	3
7. I felt that everything I did was an effort	0	1	2	3
8. I felt hopeful about the future	0	1	2	3
9. I thought my life had been a failure	0	1	2	3
10. I felt fearful	0	1	2	3
11. My sleep was restless	0	1	2	3
12. I was happy	0	1	2	3
13. I talked less than usual	0	1	2	3
14. I felt lonely	0	1	2	3
15. People were unfriendly	0	1	2	3
16. I enjoyed life	0	1	2	3
17. I had crying spells	0	1	2	3
18. I felt sad	0	1	2	3
19. I felt that people dislike me	0	1	2	3
20. I could not get "going"	0	1	2	3

6. **How much do you agree with these statements in terms of neighborhood cohesion? (Sampson, Raudenbush, & Earls, 1997)**

	Strongly disagree	Disagree	Not sure	Agree	Strongly agree
1. Residents are willing to help their neighbours	1	2	3	4	5
2. This is a close-knit and strong social inclusive neighborhood	1	2	3	4	5
3. Residents in this neighbourhood can be trusted	1	2	3	4	5
4. Residents in this neighborhood generally don't get along with each other	5	4	3	2	1
5. The relationship between residents of this neighborhood and surrounding residents (e.g. villagers) is very good	1	2	3	4	5
6. It is acceptable that outsiders share service facilities with us, such as the community clinics, clubs, and open space	1	2	3	4	5

7. **How much do you agree with the following statements? (If you do not need to work, skip to question 8)**

	Strongly disagree	Disagree	Not sure	Agree	Strongly agree
1. Living here could release my negative emotions and stress	1	2	3	4	5
2. Daily commuting has little influences on my health	1	2	3	4	5

(continued)

(continued)

	Strongly disagree	Disagree	Not sure	Agree	Strongly agree
3. I will be healthier if I travel more actively everyday	5	4	3	2	1
4. I am usually very tired after work	5	4	3	2	1
5. My job consumes me too much time, taking up my time for exercises and relax	5	4	3	2	1

8. **How much do you agree with the following statements in terms of healthy living?**

	Strongly disagree	Dis-agree	No opinion	Agree	Strongly agree
1. The green space of the neighborhood could help me to release my stress	1	2	3	4	5
2. The design and arrangement of squares increase my willing to do outdoor physical exercise	1	2	3	4	5
3. The ambient environment is good for my health	1	2	3	4	5
4. There are weak connections between the residential environment and my health	5	4	3	2	1
5. Symbolic and distinctive design of residential environment could increase the neighborhood attachment	1	2	3	4	5
6. Neighborhood attachment is good for my health	1	2	3	4	5

(continued)

(continued)

	Strongly disagree	Dis-agree	No opinion	Agree	Strongly agree
7. Communicating with neighbors could help me to improve my emotional health such as alleviating depression	1	2	3	4	5
8. Getting help from neighbors makes me feel less stressful	1	2	3	4	5
9. A good relationship with neighbors makes me feel happy	1	2	3	4	5
10. Participating in community activities is beneficial for my happiness	1	2	3	4	5
11. The poor quality of housing management often make me unhappy	5	4	3	2	1
12. Overall, living in this neighborhood is good for my health	1	2	3	4	5

9. **For general sickness, where will you go for checking up usually?**
    (1) Community clinics
    (2) Hospital in Panyu District
    (3) Hospital in urban core area
    (4) Others, please specify________________

Part 2: Residential activities

These questions are about how you traveled from place to place, including to places like work, shopping, recreation, and so on.

10. **Which means of transport do you usually use when you go to work, daily shopping and recreate (rest and exercises)? (You can choose more than one option, if you transfer.)**

	na	Walk	Cycl-ing	Bus	Sub-way	Taxi	Private car	Shuttle bus	Transit bus	Motor-taxi
Go to work	A	B	C	D	E	F	G	H	I	J
Going daily shopping	A	B	C	D	E	F	G	H	I	J
Picking up children	A	B	C	D	E	F	G	H	I	J
Going for recreation	A	B	C	D	E	F	G	H	I	J

11. **How much time do you usually spend on traveling averagely in an ordinary day (sum of travel to work, shopping and recreation)?**

Type	NA	<30 min	30–59 min	60–89 min	1.5–2 h	>2 h
Walking	0	5	4	3	2	1
Cycling	0	5	4	3	2	1
Public transportation	0	5	4	3	2	1
Private car	0	5	4	3	2	1

These questions are about the leisure activities you do during past six months, such as walking/cycling/jogging/playing basketball/swimming/dancing/go to gym, etc.

12. **What are the two most common leisure activities you participated in during the past six months? (Please specify)** (If you only do one common physical activity, then put NA in the second blank; if you do not do any leisure activities, please skip to question 17)

    (1) ______________________________

    (2) ______________________________

13. **Where do you usually do above leisure activities? (Multiple choice)**

Type	At home	In clubs of the neighborhood	Green space or squares within the neighborhood	Green or open space in Panyu district	Parks, squares, or sport venue in urban core area	Others
(1)	A	B	C	D	E	F____
(2)	A	B	C	D	E	F____

14. **During past six months, how often do you do the mentioned physical activities?**

Type	Everyday	More than 3 times a week	Once or twice a week	Once two weeks	Once/month	Less than once/month
(1)	6	5	4	3	2	1
(2)	6	5	4	3	2	1

15. **How long do you usually do the mentioned physical activity each time?**

Type	Less than 30 min	30-60 min	More than 60 min
(1)	1	2	3
(2)	1	2	3

The following section is about the interactions with your neighbors during last six months in the neighborhood.

16. **What types of informal interactions[5] have you usually engaged in during last six months? (Multiple choice)**

    (1) Talking with neighbors in public space (including open space and public facilities)
    (2) Baby-siting together in outdoor public space
    (3) Playing cards or Majiang in outdoor public space
    (4) Physical exercising with neighbors in outdoor public space
    (5) Helping (getting help) neighbors in person
    (6) Exchanging information with neighbors through virtual community (such as social software, online community forum)
    (7) Helping (or getting help) neighbors through virtual community
    (8) Others, please specify:______________________
    (9) None.

17. **What types of community activities have you usually engaged in during last six months?**

    (1) Cultural and recreation activities organized by housing manager
    (2) Activities organized by residential committee or other social organizations
    (3) Others _____________
    (4) Never attend.

18. **How do you usually participate in the housing operation affairs[6]? (Multiple choice)**

---

[5] Informal interactions, referring to unplanned activities, which is not organized by one particular organization.
For instance, bumping into a neighbor while running and errand or seeing a "regular" at the coffee shop.

[6] Housing operation affairs: affairs related to the interests of residents in daily housing operation.

(1) To be the member of home ownership association (HOA)
(2) Establish the self-organization among residents to deal with the affair
(3) Join the self-organized community initiative
(4) Appeal to or argue with the housing manager or the residential committee
(5) Discuss the affairs with neighbours face-to-face or in virtual community
(6) Pay attention to the affair but do not participate in discussion
(7) Others, please specify________________________________
(8) Never participate in any affairs.

19. **How often do you interact with your neighbors averagely during past six months? (Multiple choice)**

Type	Everyday	More than 3 times per week	Once to twice per week	Once two weeks	Once per month	Less than once per month	Never
Neighborhood interactions (such as talks, play cards, exercise)	6	5	4	3	2	1	0
Attending community activities	6	5	4	3	2	1	0
Participate in housing operation and neighborhood development affairs	6	5	4	3	2	1	0

20. **How much do you agree with these statements (focusing on residents' attitude to health-related residential behavior)?**

	Strongly disagree	Dis-agree	Not sure	Agree	Strongly agree
1. I love doing leisure activities such as the physical exercises	1	2	3	4	5
2. I had physical exercises regularly during past six months	5	4	3	2	1

(continued)

(continued)

	Strongly disagree	Dis-agree	Not sure	Agree	Strongly agree
3. Compared to other means, I prefer to walk or cycle to destinations in daily life	1	2	3	4	5
4. I like to interact with neighbors	1	2	3	4	5
5. In my spare time, I'd rather to stay at home than to interact with my neighbors	5	4	3	2	1
6. I have no interests in the community activities	5	4	3	2	1
7. I am not satisfied with the housing management	5	4	3	2	1
8. I do not care with the affairs happened in the neighborhood	5	4	3	2	1
9. There are a lot of neighbors concerning housing operation affairs	1	2	3	4	5
10. There is no easy way for residents to take part in housing operation affairs	5	4	3	2	1

Part 3: Residents' perception of the residential environment

21. **Within the neighbourhood, how do you perceive the convenience of walking to the following places?**

Within the neighbourhood, from your home						
	Very incon-venient	Incon-venient	Neutral	Con-venient	Very con-venient	No opinion
To subway stations	1	2	3	4	5	0
To bus stops	1	2	3	4	5	0
To market or shops	1	2	3	4	5	0

(continued)

(continued)

Within the neighbourhood, from your home						
	Very incon-venient	Incon-venient	Neutral	Con-venient	Very con-venient	No opinion
To recreational facilities (Club/park/squares)	1	2	3	4	5	0
To local schools or kindergarten	1	2	3	4	5	0
To community clinics or pharmacies	1	2	3	4	5	0
To community service center (bank or offices of housing manager)	1	2	3	4	5	0

22. **How much do you agree with these statements in terms of perceived residential environment?**

	Strongly disagree	Dis-agree	Not sure	Agree	Strongly agree
1. I feel the population density is all right, neither too crowd or too deserted	1	2	3	4	5
2. The neighborhood is safe from crime	1	2	3	4	5
3. The green space (including trees, grass, etc.) is not enough in this neighborhood	5	4	3	2	1
4. Within the neighborhood, the automobile traffic (amount and speed) influences my willing of walking to my destinations	5	4	3	2	1

(continued)

(continued)

	Strongly disagree	Dis-agree	Not sure	Agree	Strongly agree
5. It is convenience to walk from my home to the service facilities within the neighborhood	1	2	3	4	5
6. The walking surface (pavement, street lamp, trees) is suitable for walking	1	2	3	4	5
7. There are enough outdoor open space (square, walking path, etc.) to do exercises	1	2	3	4	5
8. There are not enough fitness facilities (playing court, swimming pool, fitness equipment/club houses) within the neighborhood	5	4	3	2	1
9. I am satisfied with the quality of the open space and fitness facilities for physical exercises	1	2	3	4	5
10. There are enough social space within the neighborhood for neighbors interactions and community activities	1	2	3	4	5
11. There are not enough health facilities, which could not satisfy my demand	5	4	3	2	1
12. There are not enough commercial facilities, which could not satisfy my demand	5	4	3	2	1

(continued)

(continued)

	Strongly disagree	Dis-agree	Not sure	Agree	Strongly agree
13. The schools within this neighborhood could not satisfy my demand	5	4	3	2	1
14. The appearance of the landscape and architecture are the symbols of this neighborhood	1	2	3	4	5
15. The reputation of the developer is the symbol of this neighborhood	1	2	3	4	5
16. The gate and signs of this neighborhood create a sense of belonging	1	2	3	4	5
17. I have the feeling of attachment living in this neighborhood	1	2	3	4	5

Thanks very much for your participation!

Appendix D
Interview Questions and Interviewee List

1. **Questions for residents**

   (1) Physical health

   - How do you evaluate your physical health condition?
   - Do you control your weight intentionally? What are the major factors influencing your physical health in your opinion?
   - Is there any effect of individual socio-demographic characteristics such as age and family structure on your physical health condition, and why?
   - How does suburban living influence your lifestyle and health condition?

   (2) Mental health

   - Is there any influence of your job on your moods and emotional wellbeing, and why?
   - What are the major factors influencing your mental health in your opinion?

   (3) Physical activity pattern

   - How long do you walk per day and what are the effects of walking on your health condition?
   - Which travel mode do you choose for commuting and how does it affect your health condition? How do you think the effects of commuting by car and by public transportation on your health, and why?
   - Do you attend physical leisure activities regularly? What are the effects of conducting physical leisure activates on your health mentally and physically? What are the factors influencing the frequency of your attendance?

   (4) Neighbors' interaction and participation

T. Zhang, *Towards Healthy Settlements*, Urban Sustainability,
https://doi.org/10.1007/978-981-97-1207-6

- Do you like to interact with your neighbors? What forms of interaction and what frequency do you prefer?
- How do the neighbors' interactions influence your emotional wellbeing? Is there a sense of belonging and why?
- Are you satisfied with the housing operation and why? Do you like to participate in the operation affairs? Is there any effect on your health?

(5) Perception on residential environment

- How do you like the physical living environment in this neighborhood in general? What kind of design do you like for wellbeing?
- How do you assess the green and open space, fitness facilities in this neighborhood? What are the effects on your health?
- Are you satisfied with the education and commercial facilities and why? What are the effects on your health?

2. **Questions for officials**

- What are the initiatives for the local government to develop the suburban housing estates?
- How do the government intervene the commodity housing development in suburbs in Guangzhou? What are the significant aspects that should be controlled by the government?
- What are the relations between the local government and developers in housing development process?
- What is the effectiveness of the regulatory plan in term of the guarantee the public interests during commodity housing development?
- What the major issues and their causes for the lacks of public service facilities in the suburban housing estates?
- What are your opinions on the healthy urban planning and design? How do you evaluate the three cases of LNT, AG and GV?

3. **Question for developers**

- How do you decide the development strategy for the commodity housing estates in suburbs? What is the importance of the target population?
- What are the most important elements to be considered in the decision-making during the whole process of housing development?
- What factors affect the quality of physical environment, such as the provision of the service facilities?
- What are the effects of regulatory plan on the housing project? Is there any negotiation space for the developer and the government?
- What are your opinions on health performance of the housing estates?

4. **Question for planers**

- How do you think the planning and design affect human health?
- What is the core value for you to plan and design a housing project?

- How do planner, developers, and the government interact with each other in the planning and design process for the housing project? What are the conflicts?
- What are the distinctive characteristics for planning and design the suburban housing estates?
- Is there any technical tools to promote neighbor's interactions and how? How do you consider the residents' behaviors in the planning and design?

5. **Question for housing managers**

- What are the major difficulties in terms of the housing operation in this neighborhood?
- Do you have any conflicts with residents in the housing operation, and why? What is the relationship between the residents' committee and housing manager?
- How do you evaluate the housing operation in terms of the facilities maintenance and community activity organization?

6. **Question for residents' committee**

- What are the shortages of this neighborhood in terms of promoting residents' health?
- How do the housing manager and residents interact with each other? Are there any conflicts and how?
- Is there any community activities organized by the residents' committee? What is the effectiveness in terms of nurturing the sense of community?
- What are the major difficulties in terms of promoting healthy lifestyle from the perspective of housing operation?

Interviewee List

Type	Number	Basic characters
Resident (LNT)	1	LNT, female, young, single, homeowner
	2	LNT, female, middle age, have a child, homeowner
	3	LNT, male, middle age, have a child, homeowner
	4	LNT, female, middle age, have a child, tenant
	5	LNT, female, middle age, have a child, tenant
	6	LNT, male, middle age, have a child, tenant
	7	LNT, female, young, single, living with parents
	8	LNT, female, middle age, have a child, homeowner
	9	LNT, male, young, single, homeowner
Resident (AG)	1	AG, female, middle age, have a child, homeowner
	2	AG, female, elderly, three-generation, homeowner
	3	AG, male, young, single, homeowner
	4	AG, female, young, have a child, homeowner
	5	AG, male, elderly, three-generation, homeowner

(continued)

(continued)

Type	Number	Basic characters
	6	AG, male, elderly, one generation, homeowner
	7	AG, male, young, have a child, homeowner
	8	AG, female, young, one generation, homeowner
Resident (GV)	1	GV, female, elderly, three-generation, homeower
	2	GV, female, middle age, have two children, homeowner
	3	GV, male, young, three-generation, homeowner
	4	GV, female, elderly, three-generation, homeowner
	5	GV, female, elderly, three-generation, homeowner
	6	GV, female, elderly, one generation, homeowner
Developer	A	Working in Vanke Property
	B	Working in Hopefluent Group Holding Limited
	C	Working in R&F Property
	D	Working in Agile Garden Property
Official	A	Working in the municipal government
	B	Working in the Panyu district government
Planner	A, B, C, D, E, F	Working in the planning institute for CAPOL, whose specialty is housing planning and design
Housing manager	A	Working in the property company for LNT
	B	Working in the property company for AG
	C	Working in the property company for GV
Residents' committee	A	Working in the residents' committee of LNT
	B	Working in the residents' committee of AG

GPSR Compliance

The European Union's (EU) General Product Safety Regulation (GPSR) is a set of rules that requires consumer products to be safe and our obligations to ensure this.

If you have any concerns about our products, you can contact us on ProductSafety@springernature.com

In case Publisher is established outside the EU, the EU authorized representative is:

Springer Nature Customer Service Center GmbH
Europaplatz 3
69115 Heidelberg, Germany

Batch number: 10726135

Printed by Printforce, the Netherlands